# WORKSHOP PHYSICS® ACTIVITY GUIDE
# THIRD EDITION

**Activity-Based Learning**

## THE CORE VOLUME
## WITH
## MODULE 1: MECHANICS I

Kinematics and Newtonian Dynamics
(Units 1-7)

## PRISCILLA W. LAWS,
## DAVID P. JACKSON AND BRETT J. PEARSON

| DIRECTOR | Andrea Pellerito |
|---|---|
| ASSISTANT EDITOR | Samantha Hart |
| SENIOR MANAGING EDITOR | Judy Howarth |
| PRODUCTION EDITOR | Mahalakshmi Babu |
| COVER PHOTO CREDIT | © photo/Getty Images |

This book was set in 10.5/12.5 Aldus LT Std Roman by Straive™.

Founded in 1807, John Wiley & Sons, Inc. has been a valued source of knowledge and understanding for more than 200 years, helping people around the world meet their needs and fulfill their aspirations. Our company is built on a foundation of principles that include responsibility to the communities we serve and where we live and work. In 2008, we launched a Corporate Citizenship Initiative, a global effort to address the environmental, social, economic, and ethical challenges we face in our business. Among the issues we are addressing are carbon impact, paper specifications and procurement, ethical conduct within our business and among our vendors, and community and charitable support. For more information, please visit our website: www.wiley.com/go/citizenship.

ISBN: 978-1-119-85652-8 (PBK)

*Library of Congress Cataloging-in-Publication Data*

LCCN 2023016539

The inside back cover will contain printing identification and country of origin if omitted from this page. In addition, if the ISBN on the back cover differs from the ISBN on this page, the one on the back cover is correct.

SKY10049270_062123

# CONTENTS

# PREFACE

*The principle of science, the definition almost, is the following: The test of all knowledge is experiment....But what is the source of knowledge? Where do the laws that are to be tested come from? Experiment, itself, helps to produce these laws, in the sense that it gives us hints. But also needed is imagination to create from these hints the great generalizations—to guess at the wonderful, simple, but very strange patterns beneath them all, and then to experiment to check again whether we have made the right guess.*

—Richard Feynman, *The Feynman Lectures on Physics*

This is the third edition of the Activity Guide developed as part of the Workshop Physics Project. Although this Guide contains text material and experiments, it is neither a textbook nor a laboratory manual. It is a student workbook designed to serve as the foundation for a two-semester, calculus-based introductory physics course sequence that is student-centered and focuses on hands-on learning. The activities have been designed using the outcomes of physics education research and honed through years of classroom testing at Dickinson College. The Guide consists of 28 units that interweave written descriptions with activities that involve predictions, qualitative observations, explanations, equation derivations, mathematical modeling, quantitative experimentation, and problem solving. Throughout these units, students make use of a flexible set of computer-based data-acquisition tools to record, display, and analyze data, as well as to develop mathematical models of various physical phenomena.

The Activity Guide represents a philosophical and pedagogical departure from traditional physics instruction. Students who study science in lecture-based courses are presented with definitions and theoretical principles. They are then asked to apply this knowledge to the solution of textbook problems and the completion of equation-verification experiments. A major objective of Workshop Physics is to help students understand the basis of knowledge in physics as a subtle interplay between observations, experiments, definitions, mathematical descriptions, and the construction of theoretical models. Instead of spending time in lectures, students in Workshop Physics make predictions and observations, do guided derivations, and learn to use computer tools to develop mathematical models of phenomena they are observing firsthand.

There are several reasons for emphasizing the processes of scientific investigation and the development of investigative skills. First, the majority of students enrolled in introductory physics courses at the high school and college levels do not have sufficient concrete experience with physical phenomena to fully comprehend the theories and mathematical derivations presented in lectures. Second, the current body of physics knowledge is truly overwhelming, and the traditional lecture method often results in trying to cover too many topics, which can lead to rote memorization on the part of the students. We believe that the only viable approach is to help students master the fundamentals and develop strategies for learning other topics independently. Finally, through many years of student evaluations we have found that most students prefer an active method of learning because it provides an environment in which asking questions and trying to clarify their understanding is encouraged.

## USING THE ACTIVITY GUIDE IN DIFFERENT INSTRUCTIONAL SETTINGS

This Activity Guide was originally designed to be used in relatively small classes in an instructional setting that combines laboratory and computer activities with discussion. Students work in collaborative groups of 2, 3, or 4 depending on the nature of each activity. Most of the activities in this Guide were tested and refined over an eight-year period at Dickinson College in a workshop environment, where sections of up to 24 students met three times a week for 2-hour class sessions. Over the years, this workshop style has been emulated at other institutions and expanded to class sizes of 50 students or more. In addition, many instructors have successfully adapted the activities for use in algebra-based courses, including courses at the high-school level. With careful planning, this Guide can even be adapted for use in large university settings. For example, many activities can be refashioned as a series of interactive lecture-demonstrations, and most can be used with little modification in smaller tutorial sessions. Thus, we believe the Activity Guide can be used effectively in almost any educational setting, particularly when in the hands of a motivated instructor.

## TOPICS COVERED

To accommodate the time required for active, hands-on learning, there are slightly fewer topics discussed in this Guide compared to a traditional lecture course. We have retained topics that are most often covered in conventional courses: Newtonian mechanics, thermodynamics, and electricity and magnetism (including circuits), and eliminated topics we felt were less essential (or often covered in later courses): fluids, waves, and optics.

Although the coverage of traditional content has been reduced somewhat, we include some topics that are not typically treated in conventional introductory physics courses: Unit 15 on Oscillations, Determinism, and Chaos; Unit 25 on Electronics; and Unit 28 on Radioactivity and Radon Monitoring.

This Activity Guide is distributed in four different modules:

Module 1: Mechanics I
Kinematics and Newtonian Dynamics (Units 1–7)
ISBN: 9781119856504 (epub); 9781119856528 (print)

Module 2: Mechanics II
Momentum, Energy, Rotational and Harmonic Motion, and Chaos (Units 8–15)
ISBN: 9781119856535 (epub); 9781119856559 (print)

Module 3: Heat, Temperature, and Nuclear Radiation
Thermodynamics, Kinetic Theory, Heat Engines, Nuclear Decay, and Radon Monitoring (Units 16–18 and 28)
ISBN: 9781119856566 (epub); 9781119856580 (print)

Module 4: Electricity and Magnetism
Electrostatics, DC Circuits, Electronics, and Magnetism (Units 19–27)
ISBN: 9781119856597 (epub); 9781119856610 (print)

The Complete Set: Modules 1–4 (Units 1–28)
ISBN: 9781119856375 (epub); 9781119856498 (print)

A typical year-long introductory physics course would use Modules 1 and 2 in the first semester and Modules 3 and 4 in the second semester, with a pace of approximately one unit per week. As with most introductory physics courses, the concepts build on each other, so we recommend covering units in numerical order, at least up through Unit 11. At that point, it's possible to skip around a little more easily (though most of the units in Module 4 should probably be covered in order).

## COMPUTER-BASED DATA-ACQUISITION SYSTEMS

Computer-based data-acquisition systems are used extensively throughout the Activity Guide for the collection, analysis, and real-time display of data. Indeed, this is one feature of Workshop Physics that is particularly powerful. The use of real-world data and graphical representations provide an immediate picture of how a physical quantity, such as an object's position or temperature, changes over time. In fact, several sensors are necessary to complete the recommended activities, including an ultrasonic motion sensor, a force sensor, a temperature sensor, a pressure sensor, a voltage sensor, a magnetic field sensor, a rotary motion sensor, and a radiation sensor. Ideally, a computer will be available for every two students in a class. However, if fewer computers are available, students can work in larger groups or participate in interactive demonstrations provided by the instructor.

As of the writing of this edition, there are several companies that produce measurement sensors and data-acquisition systems. We briefly mention two of the more popular providers: *Vernier Science Education* (www.vernier.com) and *PASCO Scientific* (www.pasco.com). Each system typically involves software running on the computer, a wide variety of associated sensors, and a method for communicating between the sensor and computer. This communication can occur through an electronic interface device connecting the computer to the sensor, or it can be done wirelessly from the sensor to the computer with no intermediate device. Detailed information regarding the setup for each type of system and the use of the associated software are included in the system manuals and software help files.

In our classrooms, we generally use wired sensors and software from *Vernier Science Education*. As part of the software distribution, *Vernier* provides activity-specific files and templates for Workshop Physics. There are a few times in the activities when we refer to these files or to some aspect of the software, but we do our best to keep software-specific language to a minimum because we realize many instructors will be using different systems. The template files generally only pre-format the data collection and graphs, so they are not technically required. Also, because computer systems and hardware will continue to change in the future, instructors may need to provide a few basic instructions to get students going. Our experience is that after a couple tips to get started, students are very adept at figuring out how to use the data collection systems.

## OTHER SOFTWARE

Analyzing motion in two dimensions can be greatly simplified by using video analysis software, and we make use of such software several times in the Activity Guide. Using such a program, position-time data can be quickly collected simply by clicking on the position of an object in each frame of a video. Most video analysis programs will also produce velocity and acceleration graphs and contain built in curve fitting (and mathematical modeling) routines as well. Both *Vernier Science Education* and *PASCO Scientific* offer video analysis tools; alternatively, there are several video analysis programs that are available as freeware.

We also note that two of our colleagues at Dickinson College, Lars English and Windsor Morgan, have developed a suite of simulations using the software package *VPython*. The simulations are designed to supplement materials for the introductory course and nicely complement the Workshop Physics Activity Guides. The materials are available as a free e-book entitled *VPython for Introductory Mechanics* by W. A. Morgan and L. Q. English.

## OTHER TRADITIONAL PHYSICS EQUIPMENT

Because the investment in computer tools and sensors is substantial, we have tried to use standard physics equipment and inexpensive items that can be acquired locally for most of the activities. We assume that instructors will have access to basic equipment, including rods, clamps, metersticks, masses, stopwatches, scales, containers, rubber stoppers, etc. However, it should be noted that we use low-friction carts and tracks extensively when studying mechanics.

## ASSIGNMENTS AND TEXTBOOK READINGS

After completing a set of activities, it is helpful for students to reinforce what they have learned by doing textbook readings and homework assignments. Although a traditional introductory physics textbook is not necessary when using the Activity Guide, we believe that such a resource is beneficial to students, particularly when seeking additional detail and clarification. In addition, such a resource provides a good source of homework problems for students. It is important that some of the homework problems focus on the types of activities and skills that students are learning in class. Thus, in addition to "traditional" textbook problems, instructors should assign problems that focus on a student's conceptual understanding of the material or their ability to analyze experimental data. Many textbooks and online homework systems now include these types of questions.

## UPDATES TO THE THIRD EDITION

This third edition of the Activity Guide represents a significant revision. While the overall structure of the Activity Guide remains largely intact, we have tried to link the Units together in a more seamless and coherent manner. As part of this process, we now emphasize vectors from the very beginning and use vector notation throughout. Becoming comfortable with vectors and vector notation is a significant stumbling block for students, and we believe that repetition is the best way to overcome this barrier. We have also attempted to place a greater emphasis on some of the more fundamental aspects of physics, particularly Newton's second law, the Momentum Principle, and the Work-Energy Principle (and to a lesser extent the Rotational Momentum Principle). By emphasizing these fundamentals, especially surrounding concepts of energy, we hope that students will gain a more consistent and methodical understanding of, for example, how mechanics and thermodynamics are intimately connected.

Lastly, we have added new *Problem-Solving* sections throughout the Activity Guide that are typically located at the end of a Unit. The activities are designed to give students practice with more challenging problems that they can tackle in a group setting (ideally when an instructor is available to provide support). We believe such practice and support will ultimately result in more skilled and confident students. In addition, the *Problem Solving* sections can be collected and graded to provide additional, regular feedback to students.

## ACKNOWLEDGEMENTS

### ACKNOWLEDGEMENTS TO THE THIRD EDITION

The new co-authors (DPJ and BJP) are deeply indebted to Priscilla Laws for her longstanding work in physics education and the development of the Workshop Physics Activity Guides. Along with her numerous colleagues and collaborators, she helped pioneer the workshop approach at a time when it ran counter to the prevailing wisdom on how to teach physics. Her perseverance in developing not only the curriculum, but also many of the necessary computer-based tools, is what made Workshop Physics a success. We are honored to contribute to the continued development of the Activity Guide.

We are grateful to our colleagues and visitors in the Department of Physics and Astronomy at Dickinson College for their conversations and input, including Robert Boyle, Krsna Dev, Lars English, Catrina Hamilton-Drager, Windsor Morgan, and Hans Pfister. Jonathan Barrick helped create the experimental set-ups for many of the new activities in the Guide. The revised materials were undoubtedly improved by the students and teaching assistants in our introductory physics classes; their patience and willingness to contribute during the development phase is very much appreciated.

We also wish to thank the following individuals whose insights and feedback helped to improve the revised edition: David Baker (Austin College), Randy Booker (University of North Carolina at Asheville), Christopher Cline and Julia Kamenetzky (Westminster College), Danielle McDermott (Los Alamos National Laboratory), and Jeff Morgan (University of Northern Iowa).

Dickinson College continues to be supportive of the Workshop Physics program. It is satisfying to work at an institution that values pedagogical development and the student learning experience. We also thank Wiley for their long-standing work on the Workshop Physics Activity Guide and related materials. In particular, Jennifer Yee, Judy Howarth, and Samantha Hart have been extremely helpful (and patient) in shepherding the third edition through the publication process and kept us on track during the inevitable hitches and delays.

David Jackson and Brett Pearson
Department of Physics and Astronomy
Dickinson College
Carlisle, PA
October 2022

### ORIGINAL ACKNOWLEDGEMENTS

All of us who were involved with this project owe a debt of gratitude to the Physical Science Study Committee for its pioneering work in the revitalization of introductory physics courses. Two individuals whose approach to physics teaching became popular in the 1960s deserve special mention for their insights into student learning difficulties—Robert Karplus of UC Berkeley and Eric Rogers of Princeton University. In addition, the work of Arnold Arons and Lillian McDermott of the University of Washington have provided inspiration for this work.

During the past eight years many people have contributed to the development of the Workshop Physics Project and this Activity Guide. First and foremost are the group of contributing authors: Robert Boyle (Units 16–18), Patrick Cooney (Unit 15), Kenneth Laws (Unit 25), John Luetzelschwab (Units 6–13 and 22–24), David Sokoloff (Units 3–7, 14, 16, 17, and 22–24), and Ronald Thornton (Units 3–7, 14, 16, and 17).

The following colleagues who have taught sections of the Workshop Physics courses at Dickinson College have contributed their insights based on the wisdom of experience. They include: Robert Boyle, Kerry Browne, David Jackson, Lars English, John Luetzelschwab, Windsor Morgan, Hans Pfister, Guy Vandegrift, and Neil Wolf. Hans Pfister deserves special mention for the design of kinesthetic apparatus for the Workshop Physics courses.

In addition, several sabbatical visitors have helped in the development of activities including Mary Brown from Dothan College, Desmond Penny from Southern Utah State College, and V. S. Rao from Memorial University in St. John's, Newfoundland.

The activities could not have been tested and refined without the work of several student generations of equipment managers and summer interns who helped during the early years of testing. They are Jennifer Atkins, Christopher Boswell, Joshua Clapper, Catherine Crosby, Ryan Davis, David Diduk, Christopher Eckert, Amy Filbin, Jake Hopkins, Michelle Lang, Mark Luetzelschwab, Despina Papazisis, Alison Sherwin, and Jeremiah Williams. I am also grateful to the 70 or so student assistants and graders and the approximately 700 students who have survived the Workshop courses as we tested and retested various activities.

Several Dickinson physics majors and project associates have developed software or software tools that have been used in the program including Grant Braught, David Egolf, Mike King, Sean LaShell, Mark Luetzelschwab, Brock Miller, and Phillip Williams. Several individuals in the Tufts University Center for Science and Mathematics Teaching who have rewritten early versions of computer-based laboratory software have redesigned portions of their software to meet our needs including Stephen Beardslee, Lars Travers, and Ronald Thornton.

The insights of colleagues from other departments and institutions have tested workshop activities or developed pedagogical approaches that have been helpful in the refinement of this Activity Guide. These colleagues are Nancy Baxter-Hastings (Dickinson College Department of Mathematics), Gerald Hart and Roger Sipson (Moorhead State University), Robert Morse (St. Alban's School), E. F. Redish and Jeffrey Saul (University of Maryland), Mark Schneider (Grinnell College), Robert Teese (Muskingham College), Maxine Willis (Gettysburg High School), William Welch (Carroll College), and Jack Wilson (Rensselaer Polytechnic Institute). Early adopters who have made contributions include Mary Fehrs and Juliet Brosing at Pacific University, Bruce Callen at Drury College, Ted Hodapp at Hamline College, Jim Holliday at John Brown University, Bill Warren at Lord Fairfax Community College, Bill Wehrbein at Nebraska Wesleyan, and Maxine Willis at Gettysburg High School.

Several administrators at Dickinson College have arranged for financial support for purchasing equipment, remodeling our classroom and equipment storage areas, and providing facilities for project staff. These individuals include President A. Lee Fritschler, Deans George Allan and Margaret Garrett, the treasurer, Michael Britton, and grants officer Christina Van Buskirk.

Individuals from the commercial sector have helped with the design, production, and distribution of hardware, software, and apparatus needed for the activities in this guide. They include: David and Christine Vernier of Vernier Software and Technology, Paul Stokstad and David Griffith of PASCO Scientific, Rudolph Graf of Science Source, and Ron and Wendy Budworth of Transpacific Computer Company.

Workshop Physics Project support staff who have helped with the production of this Activity Guide include Susan Greenbaum, Gail Oliver, Susan Rogers, Pam Rosborough, Sara Settlemyer, Virginia Trumbauer, Maurinda Wingard. Kim Banister, Erston Barnhart, Kevin Laws, Virginia Jackson, and Noel Pixley have helped with the artwork. Wiley editors Clifford Mills and Stuart Johnson with the help of Katharine Rubin and Geraldine Osnato have coordinated the Activity Guide production effort.

Major support for this work was provided by the Fund for Improvement of Postsecondary Education (Grants #G008642146 and #P116B90692-90) and the National Science Foundation (Grants #USE-9150589, #USE-9153725, #DUE-9451287, and #DUE-9455561). Project Officers who have provided administrative and moral support for this project are Rusty Garth, Brian Lekander, and Dora Marcus from FIPSE, and J. D. Garcia, Ruth Howes, Kenneth Krane, and Duncan McBride from NSF.

Priscilla Laws
Department of Physics and Astronomy
Dickinson College
Carlisle, PA
January 2004
On behalf of contributing authors Robert Boyle, Patrick Cooney,
Kenneth Laws, John Luetzelschwab, David Sokoloff, and Ronald Thornton

# UNIT 1: OUR PLACE IN THE UNIVERSE

*This stunning photo shows Earth rising above the lunar surface and is composed from a series of images taken from NASA's Lunar Reconnaissance Orbiter on October 12, 2009 (Image Credit: NASA/Goddard/Arizona State University). Seeing Earth from this perspective shows our planet floating in a vast and unimaginably large space we call the Universe.*

# UNIT 1: OUR PLACE IN THE UNIVERSE

## OBJECTIVES

1. To become familiar with the approach of *Workshop Physics*.

2. To get acquainted with making *estimates* when a measurement is impractical.

3. To understand scientific notation and significant figures, as well as how to covert between different sets of units.

4. To learn about *vectors* and coordinate systems.

## 1.1   OVERVIEW

The purpose of this unit is to introduce several important concepts and tools that will be useful during the rest of this course (and hopefully elsewhere!). We do so in the context of our place in the Universe. Throughout the vast majority of human history, we were effectively confined to the surface of Earth. People could climb mountains to reach higher elevations, but they were still "on Earth." The development of powered aircraft in the early 1900s allowed humans to leave the surface of Earth for the first time. Roughly 50 years later, humans left Earth behind by reaching what is typically referred to as "outer space." Going beyond Earth, we end this unit by investigating Earth's place in our solar system, and from that, our solar system's place in the Universe.

    We begin with a discussion of units, conversions, and scientific notation, and then move on to describe *vectors*. Vectors, which have a direction as well as a magnitude, are essential for describing concepts such as position, velocity, acceleration, and force. Thus, understanding the basics of vectors, including how they can be added, subtracted, and multiplied, is critical in the study of physics.

## ESTIMATES, UNITS, CONVERSIONS, AND SCALES

### 1.2  ESTIMATES AND UNIT CONVERSIONS

ansonsaw/Getty Images

Thousands of jet airplanes circle Earth every day. In fact, aircraft are so ubiquitous in our society that at any given time there's a good chance you can look up in the sky and see an airplane flying high above you. Obviously, such an airplane will look quite small relative to its actual size, a byproduct of how far away it is. But just how far away is it?

For the activities in this section, you will need the following equipment:

- 1 tennis ball

---

#### 1.2.1.  Activity: Estimating Heights

**a.** Go outside with your group and try throwing a tennis ball straight up as high as you can, being careful not to damage anything (or anyone!). Each of you should make two or three attempts. The goal here is to *estimate* how high you can throw the ball. Explain below how you came up with your estimate.

**b.** Now let's try something a little more challenging. Without looking it up, try to determine the approximate cruising altitude of a large, jet airplane. Talk it over with your group and come up with an estimate based on your experiences and what you've observed when seeing airplanes in the sky. (If you don't know where to begin, start small and work your way up. For example, could it be 100 ft? 1,000 ft? 10,000 ft? etc.)

---

#### Estimations Are Approximate Answers

If you have a lot of experience with airplanes, the previous question might seem easy. On the other hand, if you've never been on a commercial flight before, you might not have any idea how to make such an estimate. One way to begin is to start small and work your way up. For example, it is probably obvious to you that an airplane flies at an altitude greater than 100 ft, since it is not uncommon for a tree to reach a height of 100 ft. How about 1,000 ft? Again, if you think about it for a while, you might realize that some skyscrapers are taller than 1,000 ft. How about 10,000 ft? At this point, most people start to feel less certain. Clearly, there are mountains that reach this height, so it might seem likely that airplanes would need to fly higher, but it's not obvious. How about 100,000 ft? This height is well beyond the world's highest mountain so this altitude might start to seem a little extreme; after all, there's no obvious reason why a plane would need to be so high. How about 1,000,000 ft? By now, most people start to feel that this is much too high.

So how do we determine a final answer? By narrowing down the range and choosing something "in the middle." If 10,000 ft seems possible, but maybe a bit on the low side, and if 1,000,000 ft seems way too high, then a reasonable guess might be 100,000 ft, or perhaps a bit less since 10,000 ft didn't seem completely unreasonable. Notice that we didn't pick the point that lies halfway between 10,000 ft and 1,000,000 ft; instead, we kept increasing the altitude by a factor of 10 and picked the one between what we knew to be too low (10,000 ft) and what we knew to be too high (1,000,000 ft).[1] Obviously, we can't be too confident in this answer, but it certainly seems plausible. We will return to this question shortly, but first let's try another estimation problem.

### 1.2.2. Activity: Distance Around Earth

Working with your group, try to estimate the distance around Earth in miles. If this sounds impossibly difficult, here are some things to think about that might be helpful: flying in an airplane; long car rides; imagining a globe; considering time zones; and the size of your home country. Spend about five minutes on this question and try to arrive at an estimate that the entire group agrees is reasonable. Provide your reasoning and write down a distance you are confident is too small, a distance you are confident is too large, and your final estimate.

You might be surprised to see that most groups come up with estimates that are somewhat similar, even though they might have used different approaches to the problem. Obviously, the answers won't be exactly the same, but they are usually "in the same ballpark." We say they have the same *order of magnitude*. They may differ by a factor of 2 or 3, but they probably don't differ by a factor of 10.[2] One of the keys to making reasonable estimates is to try to relate some experience you have had, such as driving across a state or flying across the country, to the question being posed. Of course, this is often easier said than done.

Let's now return to the cruising altitude of airplanes. It is not uncommon for students to have estimates ranging from 10,000 ft up to 100,000 ft, and all these answers are reasonable as an estimate. Notice that 10,000 ft and 100,000 ft

---

[1] The altitude halfway between 10,000 ft and 1,000,000 ft is 505,000 ft, a point referred to as the *arithmetic* mean (*add* the two numbers and divide by two). But notice that 500,000 ft is 50 times larger than 10,000 ft, whereas 1,000,000 ft is only two times larger than 500,000 ft. In this sense, 500,000 ft is much "closer" to 1,000,000 ft than it is to 10,000 ft. Thus, when trying to estimate something we don't know, we typically take the *geometric* (or *multiplicative*) mean (*multiply* the two numbers and take the square root).

[2] A factor of 10 is considered to be an *order of magnitude*, while two factors of 10, or a factor of $10^2 = 100$, is considered two orders of magnitude (and so on).

differ by an order of magnitude, so any estimate that lies between these values is within one order of magnitude. It turns out that large, jet airplanes typically fly at an altitude of 30,000–40,000 ft above sea level. Thus, when you are asked to estimate the *approximate* cruising altitude, there is no single correct answer. Having said that, some answers are better than others. For example, 35,000 ft and 40,000 ft are both perfectly acceptable answers; one is not any better than the other, as they both give an accurate estimate of the cruising altitude. Similarly, the answers of 25,000 ft and 45,000 ft, while outside the typical range for most commercial jets, are also fine if you are just looking for an approximate answer. On the other hand, an answer of 100 ft would clearly be incorrect.

The point here is that it's the *approximate* range that's important, not the specific answer. In fact, an answer such as 34,476 ft is not really an acceptable response to the question, as this answer implies a very precise knowledge—down to one foot—of how high a typical airplane flies. This is not to say that no plane has ever flown at this altitude, but that this answer provides a very specific altitude, whereas the question asks only for an approximate answer. (We will come back to the idea of implied precision shortly.)

In this course, we will often do estimates such as this, especially when we are making a prediction about something. These estimates are sometimes referred to as *order-of-magnitude approximations*. When making such approximations, you have some flexibility in choosing your answer. In fact, depending on the calculation you are doing, it might be advantageous to choose one estimate over another. For example, suppose you needed to divide your answer by six. In this case, it might be a good idea to use 36,000 ft as an estimate because this value is easily divisible by six without using a calculator. On the other hand, if you needed to divide by five instead of six, then 35,000 ft would be easier to work with.

---

### 1.2.3. Activity: Altitude in Miles

**a.** In this activity, we want you to convert your answer from Activity 1.2.1. (b) from feet to miles. To do so, you need to know that there are 5,280 ft in 1 mile (this is a useful conversion to remember). Begin by estimating the altitude in miles *without* using a calculator. To do so, you will need to make some approximations, so be sure to show your work. (If your previous estimate happened to be well outside the typical cruising altitude, you should choose a more appropriate value for this calculation.)

**b.** Now assume you've been told that a particular airplane is flying at an altitude of 32,500 ft and you're asked to determine the exact altitude in miles. (In this case we are *not* asking for an estimate, so you *should* use a calculator to arrive at your answer.)

**Unit Conversions: Multiplying by One**

For the above calculation, you probably quickly figured out that you needed to *divide* your answer by 5,280 to convert from feet to miles. Formally, one can think about multiplying your answer by a "conversion factor" that is given as a fraction whose value is one. For example, for an airplane flying at an altitude of 32,500 ft, you would convert this to miles as follows:

$$32{,}500 \text{ ft} \left( \frac{1 \text{ mile}}{5{,}280 \text{ ft}} \right) \approx 6.16 \text{ miles}$$

A few things to notice:

- The units of feet in the numerator and denominator cancel, leaving only the desired units of miles.
- The numerator and the denominator in the fraction are the same type of unit (length) and both represent a distance of one mile, so the fraction has a value of one (one mile divided by 5,280 ft equals one). Therefore, you can multiply your answer by the fraction without changing its value.
- Note that the units are an essential part of the conversion factor; without them the conversion factor is no longer equal to one. (One mile divided by 5,280 feet is equal to one, but the number 1 divided by the number 5,280 is obviously *not* equal to one.)
- Because the conversion factor is equal to one, so is its reciprocal. That is, 5,280 feet divided by one mile is also equal to one. Which fraction to use will depend on whether you are converting from feet to miles or from miles to feet. The trick is to choose the conversion factor to cancel the units you don't want and leave you with the units you desire.

While this procedure may be unnecessary for such a simple conversion as above, we recommend going through this process whenever you convert units. Doing so makes the conversion process relatively simple, particularly when you need to do a series of conversions.

### 1.2.4. Activity: More Complicated Conversions

**a.** To get some experience with unit conversions, convert the speed 60 miles per hour (mph) into meters per second by using multiple (six) different conversion factors. Show explicitly which units cancel and which ones don't. **Hint**: There are 2.54 cm in one inch; we assume you know how many seconds are in a minute and how many minutes are in an hour.

**b.** Some quantities involve units that are squared or even cubed. For example, atmospheric pressure can be measured in pounds per square inch (lbs/in$^2$) or Newtons per square meter (N/m$^2$). Suppose atmospheric pressure is measured to be 14.5 lbs/in$^2$. Determine the pressure in Newtons per square meter. Note that the units of inches squared will require squaring some of the conversion factors. **Hint**: 2.2 lbs is equal to 9.8 N, the weight of 1 kg of mass.

## 1.3  SIGNIFICANT FIGURES AND SCIENTIFIC NOTATION

In Section 1.2 we briefly discussed the fact that 34,476 ft would not be an appropriate response for an estimate since it implies a very precise level of knowledge. The idea of "implied precision" (or "implied uncertainty") comes from the number of *significant figures* in the answer, and how this relates to both approximations and measurements will come up throughout this course. Although we will not be overly concerned with significant figures in this book, it is worth discussing the basics.

rookie72/Adobe Stock

### Significant Figures and Implied Uncertainty

A *significant figure* in a number is one that contributes to its known precision. This is probably easiest to think about in relation to a simple measurement. Suppose you wanted to measure the width of your calculator using a ruler. The smallest markings on a standard ruler are in millimeters, so you could certainly measure the width to the nearest millimeter, say, 78 mm. This number has two significant figures, as they both contribute to the known precision of the measurement. It is important to realize, however, that this number does not suggest that the width of the calculator is *exactly* 78 mm. Instead, it suggests that it is *closest* to 78 mm, but could still be off by some amount. If no measurement uncertainty is provided, then the *implied uncertainty* would be ±0.5 mm. You are confident it is closer to 78 mm than either 77 mm or 79 mm, but within that range you are uncertain. In other words, you are confident the width is between 77.5 mm and 78.5 mm (if you are not confident of this, then you should not have specified that the width of the calculator was 78 mm!).

Now, if you have good eyesight and a steady hand, you might be able to determine that the width is roughly halfway between 78 mm and 79 mm, in which case you could give its width as 78.5 mm. This number has three significant figures—all three of them contribute to the known precision. Again, if no uncertainty is specified, then the implied uncertainty here would be ±0.05 mm; that is, the width lies somewhere between 78.45 mm and 78.55 mm. However, when using only a ruler and your eyes for this measurement, it would not make sense to list the width as 78.52394616 mm. There is simply no basis for including any digits beyond the 5, which is already a bit uncertain. When performing calculations using a calculator, it may be tempting to write down all the digits shown on the screen. But remember that this number represents some physical quantity (e.g., the width of the calculator), and it is unrealistic to specify such a precise value.

If we return to our earlier estimate of a typical airplane cruising altitude, giving an answer of 34,476 ft implies an uncertainty of ±0.5 ft. This precision is clearly not warranted for our estimate. Moreover, such an uncertainty does not even make physical sense. For example, large airplanes are over 50 ft tall, so this uncertainly would require you to specify the exact point on the plane you are referencing. In addition, it is extremely unlikely that an airplane could keep its altitude constant to within half a foot while flying. Even if its altitude was 34,476 ft at one instant of time, small fluctuations in air temperature, pressure, winds, etc. would certainly cause it to drift up or down by more than a few feet.[3]

---

### 1.3.1. Activity: Significant Figures and Implied Uncertainty

Determine the number of significant figures and the implied uncertainty for each number as it is written.

**a.** 125

**b.** 4.7

**c.** 36,400

---

Note that it may be reasonable to specify an uncertainty that is *larger* than the implied uncertainty. As an example, consider measuring someone's height. This is often done by having the person stand against a wall and placing a pencil horizontally against the top of their head so that it extends to the wall and makes a mark. The person's height is then determined by measuring the distance from the floor to this mark on the wall. While it might be possible to measure this distance to an accuracy of, say, 1 mm, there is no guarantee that the mark on the wall corresponds perfectly to the height of the person. After all, the pencil used to make the mark on the wall might have been inadvertently angled so that the mark ended up a little higher or lower than the person's actual height. Thus, it might be reasonable to assign an uncertainty of, say, ±4 mm to the height, even though the length measurement has a precision of 1 mm. In such a situation, the height would then be reported as $1.772 \pm 0.004$ m.

---

[3] There is also the issue of the precision achievable by the airplane's *altimeter* (the device that measures altitude). Although we don't readily know how precise an altimeter can be, it seems reasonable to assume that it probably can't reliably distinguish different altitudes that are just a few feet apart.

### Scientific Notation

Sometimes the number of significant figures can be confusing or even ambiguous, especially when the numbers contain zeros, as in part (c) of the previous activity. For example, if someone reported a measurement of 1,000 cm, it's not clear whether this number has one, two, three, or four significant figures.[4] For this reason, it is useful to write numbers using *scientific notation*, where the number of significant figures is clear.

A number expressed in scientific notion is given as a number between one and ten (typically followed by one or more decimal places) multiplied by a power of 10. For example, the number 34,476 (ignoring any potential units for now) would be written in scientific notation as $34{,}476 = 3.4476 \times 10^4$. Because $10^4 = 10{,}000$, multiplying 3.4476 by $10^4$ yields the original number. You are effectively moving the decimal point four places to the right when you multiply by $10^4$. A negative exponent implies the reciprocal of the corresponding number with a positive exponent; in other words, $10^{-3} = 1/10^3$. If the exponent is negative, you move the decimal point in the opposite direction (to the left): $0.00429 = 4.29 \times 10^{-3}$.

For very large or very small numbers, scientific notation has the obvious benefit of being more compact. For example, the number fourteen billion (roughly the age of the Universe in years) is more compactly written as $1.4 \times 10^{10}$ compared to 14,000,000,000. Similarly, the number one millionth is more easily written (and understood) as $1 \times 10^{-6}$ compared to 0.000001.

Another benefit is that it is easier to multiply and divide numbers when they are expressed in scientific notation since the powers of 10 combine easily. For example, when multiplying 2 powers of 10, the exponents are added together ($10^3 \times 10^2 = 10^{3+2} = 10^5$), whereas when dividing powers of 10, the exponents are subtracted ($10^3 \div 10^2 = 10^{3-2} = 10^1$). Note that multiplication and division of powers of 10 are closely related due to the reciprocal rule: $10^3 \div 10^2 = 10^{3-2} = 10^{3+(-2)} = 10^3 \times 10^{-2}$.

Scientific notion also clearly indicates the number of significant figures and the corresponding implied precision. The basic rule in scientific notion is that you *write all the digits that are significant, including zeros*. Consider the altitude of 36,000 ft. If you write this as $3.6 \times 10^4$ ft, you are specifying two significant figures and indicating you are confident in the altitude to the thousands digit (an uncertainty of ±500 ft). If you instead write it as $3.60 \times 10^4$ ft, you are specifying three significant figures and are confident in the altitude to the hundreds digit (±50 ft). If you were only confident to the ten-thousands digit, you would want to write it with only one significant figure, as $4 \times 10^4$ ft, where we have rounded 3.6 up to 4.

---

### 1.3.2. Activity: Practice with Scientific Notation

a. Write the following two numbers in scientific notation and then multiply them together: 25,000,000 × 0.0004. Try to do this without using a calculator! Keep your answer in scientific notation. (For this part, and

---

[4] The general rule is that leading zeros (those on the left) are *not* significant, zeros between non-zero digits *are* significant, and trailing zeros (on the right) are *not* significant unless a decimal point is shown.

those that follow, you can assume there are three significant figures in each number.[5])

**b.** Make use of scientific notation to divide 160,000 by 4,000,000 (again, without using a calculator).

**c.** A light year is defined to be the *distance* traveled by light in one year. If the speed of light is 186,000 miles per second, use appropriate conversion factors to determine the number of *miles* in a light year. Begin by converting all numbers to scientific notation. (And yes, it's okay to use a calculator for this problem.)

### 1.3.3. Activity: Significant Figures and Implied Uncertainty in Scientific Notation

Determine the number of significant figures and the implied uncertainty for the numbers below.

**a.** $6.4 \times 10^3$

**b.** $6.40 \times 10^3$

**c.** $2.00 \times 10^{-2}$

---

[5] As you may know, there are rules for how to track significant figures when adding, subtracting, multiplying, and dividing numbers. However, because significant figures are not a primary component of this course, we will simply keep three significant figures in all calculations unless specified otherwise.

Being comfortable with scientific notation is extremely important, so be sure you understand the previous activities. (If you found these activities to be reasonably simple, then take comfort in the fact that this is one less thing you need to learn.)

## 1.4  SPACECRAFT, SATELLITES, AND ORBITS

In Section 1.2 we estimated the typical cruising altitude of a large jet plane. In this section, we discuss the altitudes of objects that reach "outer space" and orbit Earth. We begin by considering NASA's space shuttle program, which was active from 1981 to 2011. The shuttle itself was a reusable vehicle that was launched with the help of attached rocket boosters, orbited (circled) Earth numerous times, and then returned to the ground, landing like a traditional airplane (Fig. 1.1).

Fig. 1.1. Space shuttle Columbia lifts off from Launch Pad 39A on April 12, 1981, to begin the first shuttle mission (STS-1). Image credit: NASA.

### 1.4.1.  Activity: Space Shuttle Altitude Prediction

In Activity 1.2.1, we saw that a large, jet airplane flies at an altitude of (approximately) 6 or 7 miles above the surface of Earth. Use this knowledge to *estimate* the approximate altitude of orbit for the space shuttle. Again, don't look it up! Remember, it can be helpful to determine values that everyone agrees are too large and too small, and then trying to narrow things down from there. (If it's helpful, it usually takes about 10 minutes for the space shuttle to reach its orbital distance after takeoff.)

### Kepler's Third Law and Orbital Period

Although we won't calculate it from first principles, there is a mathematical expression known as Kepler's third law that allows us to determine the time it takes an object to make one complete orbit around a planet (or star). For an object that orbits in circular motion about a much more massive planet (assumed stationary), Kepler's third law can be written as

$$T = 2\pi \sqrt{\frac{r^3}{GM}} \tag{1.1}$$

Here, $T$ is the time (in seconds) it takes to complete one full orbit (the so-called orbital *period*), $r$ is the distance (in meters) between the center of the object and the center of the planet (the orbital *radius*), $G$ is the so-called gravitational constant ($6.67 \times 10^{-11}$ m$^3$/kg·s$^2$), and $M$ is the mass (in kilograms) of the planet. (Note that when the planet is assumed to be stationary, the period does not depend on the mass of the orbiting object.)

Kepler's third law, in the form of Eq. (1.1), provides an expression for the period $T$ in terms of the variables $r$ and $M$. Of course, we can rearrange this equation to solve for either $r$ or $M$ as needed. For example, imagine we are given the orbital period of the space shuttle as it circles Earth ($T$ in the equation above). We could then rearrange Kepler's third law to solve for the orbital radius $r$.

Recall that the orbital radius is the distance from the center of Earth to the center of the object. In other words, the orbital radius is equal to the radius $R_E$ of Earth plus the height $h$ above the surface (the object's altitude). Therefore, by solving Kepler's third law for $r$, we can then determine the shuttle's altitude as $h = r - R_E$.

Let's start by solving Kepler's third law for the orbital radius. Diving by $2\pi$ and squaring both sides leads to

$$\left(\frac{T}{2\pi}\right)^2 = \frac{r^3}{GM}$$

Multiplying both sides by $GM$ and taking the *cube* root (raising to the power of 1/3) then yields

$$r = \sqrt[3]{T^2 \left(\frac{GM}{4\pi^2}\right)} \tag{1.2}$$

We can now use this expression to find the orbital radius, and hence the altitude of the orbit.

---

### 1.4.2. Activity: Space Shuttle Altitude Calculation

**a.** Note that the factor $\left(\frac{GM}{4\pi^2}\right)$ is fixed for any object orbiting Earth (assuming the mass $M$ of Earth remains the same). Since we will perform this calculation more than once, it will be helpful to first find a value for this combination. Calculate the quantity $\left(\frac{GM}{4\pi^2}\right)$ using the value of $G$ given above and the fact that the mass of Earth is approximately $6.0 \times 10^{24}$ kg.

**b.** The space shuttle takes approximately 90 minutes to orbit Earth. Use this fact to calculate the orbital radius of the shuttle. **Hint**: Be careful with units!

**c.** Finally, use the fact that the radius of Earth is approximately $6.4 \times 10^6$ m to calculate the altitude $h$ of the orbit. Convert your final answer for the altitude into miles and compare it to both your previous estimate and to the altitude of an airplane.

**d.** Compare the altitude you just found to Earth's radius, and then make a rough scale drawing of Earth and the space shuttle in orbit to show the relative height above the surface compared to Earth's size.

---

You should have found that the space shuttle orbits at an altitude that is quite a bit larger than the altitude at which airplanes fly. However, you should also have found that, compared to the size of Earth, the space shuttle orbits *very* close to the surface! Moreover, because the space shuttle is orbiting in "outer space," we see that Earth's atmosphere barely extends above the surface of Earth (when viewed from the proper perspective), something most people find surprising.

Of course, objects can also be put into orbit at much larger distances. But due to Kepler's third law, larger orbital distances will result in longer orbital periods, as the following activity illustrates.

### 1.4.3. Activity: Geostationary Satellite Orbit

**a.** As we just learned, the space shuttle orbits Earth with a period of about 90 minutes. In other words, it takes 90 minutes for the space shuttle to completely circle Earth. Meanwhile, Earth spins once on its axis every 23 hours, 56 minutes, and 4 seconds.[6] Thus, the space shuttle orbits Earth many times in a given day and will therefore not always be visible overhead.

Now consider a communications satellite orbiting Earth. It would be convenient if such a satellite was always "visible" overhead so that radio signals or other communications would not have to track the satellite as it moves. In other words, we would like the satellite to appear motionless in the sky above. Such an orbit is called a *geostationary* orbit. To begin, determine the period, in seconds, of such a geostationary orbit.

---

[6] As you well know, a "day" on Earth is 24 hours long. But such a "day" is the time it takes Earth to spin all the way around so that the sun returns to the same location in the sky. The proper terminology for this period of time is a *solar day*. However, because Earth is also orbiting the sun as it spins on its axis, it will need to spin by slightly more than 360° in order for the sun to appear at the same location as the previous day (nearly 1° more). Because of this extra rotation, the time it takes for Earth to spin exactly once (360°) around its axis will be slightly *smaller* than 24 hours. This slightly shorter time is called a *sidereal day*.

**b.** Now use Kepler's third law to determine the *altitude* of a satellite in geostationary orbit. You will probably find it helpful to use some of your previous calculations (and don't forget about the difference between orbital radius and altitude). Express your answer in both kilometers and miles.

**c.** Make a rough scale drawing of Earth and a satellite in geostationary orbit to show the height above the surface relative to Earth's size.

---

**Earth's Moon**

Like the space shuttle or a communications satellite in geostationary orbit, the moon also orbits Earth due to their mutual gravitational attraction.

**1.4.4. Activity: The Moon's Orbit**

**a.** Use Kepler's third law to calculate the orbital radius of the moon around Earth (we'll assume the moon orbits a fixed Earth). You will need to use what you know about the moon's orbital period (don't forget to convert this to seconds) and once again make use of your previous calculations. Show and explain your work!

NASA

**b.** Again, compare the orbital radius you just found to Earth's radius. Make a scale drawing of Earth and the moon in orbit to show the approximate separation compared to Earth's size. Include a geostationary satellite and the space shuttle on this diagram as well. **Hint:** Don't make Earth too large!

---

In 1969, NASA's Apollo 11 mission landed two people on the surface of the moon for the first time. A large rocket was used to launch both the command module (*Columbia*) and the attached landing module (the *Eagle*) from Earth to orbit around the moon. *Columbia* stayed in orbit above the moon, while the *Eagle* landed on the surface and stayed for approximately 24 hours before taking off and rejoining *Columbia*. *Columbia* then flew back to Earth, re-entered the atmosphere, and splashed down in the Pacific Ocean. Hopefully, given your answer for the moon's distance from Earth, you can appreciate just what an incredible feat this was. Even more astonishing, in perhaps the world's first streaming event, a video feed of the moon landing was broadcast live to the entire world as it was happening (see Fig. 1.2)!

**Fig. 1.2.** Buzz Aldrin walks on the surface of the moon near one leg of the landing module. Image credit: NASA.

## POSITIONS, COORDINATES, AND VECTORS

### 1.5  THE SOLAR SYSTEM

Just as the moon orbits Earth, Earth and the other planets in our solar system orbit the Sun. In this section, we will work through a few activities to get a sense of scale for our solar system (well, at least parts of our solar system).

#### 1.5.1.  Activity: Earth's Orbit

Use Kepler's third law to calculate the orbital radius of Earth as it orbits the sun (recall that this radius is the center-to-center distance between Earth and the sun). Note that because Earth is orbiting the sun, you will now need to use the sun's mass of $M = 1.99 \times 10^{30}$ kg in Kepler's law. You may want to first calculate the new value of the factor $\left( \frac{GM}{4\pi^2} \right)$, as we will use it again below. State your answer in both meters and miles.

**The Astronomical Unit (AU)**

The distance found in the previous activity is quite large and somewhat awkward to use. But when discussing planets in our solar system, it is convenient to measure distances in terms of this Earth–sun separation. Thus, we define an *Astronomical Unit* (abbreviated AU) as the (approximate) distance between Earth and the sun of $1.50 \times 10^{11}$ m, or approximately 93 million miles.[7]

---

**1.5.2. Activity: The Size of the Solar System**

a.  Consider Mercury, the closet planet to the sun, with an orbital period of 88 (Earth) days. Determine the orbital radius of Mercury and convert your answer to AU.

b.  Now, repeat this procedure for Neptune, the farthest planet from the sun, and, in some sense, the "edge" of our solar system.[8] The orbital period of Neptune is 164.8 Earth years, or 60,182 Earth days.

---

Hopefully, the previous activity gives you some sense of the scale of our solar system and the vast distances and empty space between the planets and the sun. But we need not stop there. As you look up at the night sky, you will see dozens of stars (if not hundreds or even thousands, depending on how dark your sky is), each one a sun like our own.

---

**1.5.3. Activity: How Far Away Are the Stars?**

a.  Alpha Centauri is the closest star system to our solar system; it lies 4.3 light years from our sun (we determined the distance of a light year in Activity 1.3.2). Determine the distance between the sun and Alpha Centauri in AU, meters, and miles. How does this distance compare (approximately) to the size of our solar system (is it 10 times larger, 100 times larger, etc.)?

---

[7] Because Earth's orbit is not perfectly circular, one AU represents something like an average distance between the two centers. To be precise, one AU is defined to be exactly 149,595,870,700 m.

[8] In 2006, Pluto was dropped from the list of planets in our solar system and is now classified as a dwarf planet.

To complete the picture, we could go further still. All of the stars you see at night (and many more that you can't see) comprise a large grouping of stars we call the Milky Way *galaxy*. The Milky Way contains between 100 million and 400 million stars, each a sun that likely has planets orbiting around it. The size of the Milky Way galaxy is approximately 200,000 light years across, and the Milky Way is just one of at least 100 billion galaxies that are known to exist, each containing hundreds of millions of stars. The closest galaxy is the Andromeda galaxy, which lies approximately 2 million light years from the Milky Way. Interestingly, the Universe is composed mostly of empty space, but yet still includes a staggering number of stars and even more planets. It is nearly impossible to fully grasp such large numbers and the incredibly immense size of the Universe. Indeed, such an exercise is truly humbling.

## 1.6   THE POSITION OF AN OBJECT

In the previous section, we found the orbital distance between the sun and Neptune to arrive at an approximate size of our solar system. If we wanted to study the solar system in detail, it would be important to be able to locate the precise positions of all the planets. To specify these positions, we need to decide on a location from which to measure the positions of all objects, a location we call the *origin*. The choice of where to put the origin is arbitrary, but some positions are more convenient than others, and we usually try to locate the origin to make our lives as easy as possible. For example, when studying the solar system, a convenient choice is to place the origin at the location of the sun because the sun's position changes very little due to the motion of the planets (we will assume the sun doesn't move at all).

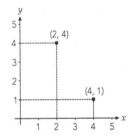

Once we've chosen the origin, we next need to specify directions so that we have a *coordinate system*. A coordinate system is (typically) a set of three, mutually perpendicular axes that correspond to the three physical dimensions of space. The most common coordinate system, and possibly the only one you have used before, is called *Cartesian coordinates* and consists of $x$, $y$, and $z$ axes all pointing outward from the origin. Cartesian coordinates are typically oriented such that the $x$-axis points in one horizontal direction (often to the right), the $y$-axis in another horizontal direction, and the $z$-axis in the vertical direction. But just as with the placement of the origin, the choice of how the axes are oriented is arbitrary, and it is wise to choose a coordinate system that simplifies the problem as much as possible. For example, if we assume the planets all orbit in a flat disc around the sun (in a single *plane*), then we only need two axes to specify the positions of the planets. By orienting our coordinate system such that the $x$ and $y$ axes are in this plane, we can ignore the $z$-direction entirely.[9]

Once we have oriented our coordinate system, the last step is to calibrate the axes—what are the units on the axes and how are they spaced? In other words, what do the tick marks on the axes represent: inches, meters, miles, or something else? Once again, the choice is arbitrary and one of convenience. For example, for the planets in the solar system, a good choice is to scale the axes so that each tick mark represents one astronomical unit (AU), or some fraction of

---

[9] We note that circular polar coordinates would probably be an even better choice of coordinate system here, but we will stick with Cartesian coordinates at this point to keep things familiar.

an AU. As we saw in the last section, an AU is a convenient unit of distance on the scale of the solar system, so this is a natural choice for our scale.

Finally, once we have chosen a scaled coordinate system, specifying the position of a planet (or any object) is as simple as specifying the *coordinates* of the object. The coordinates are a triplet of values $(x, y, z)$, where $x$ represents the position along the $x$-axis (called the $x$-coordinate), $y$ represents the position along the $y$-axis (the $y$-coordinate), and $z$ represents the position along the $z$-axis (the $z$-coordinate). If we have oriented our coordinate system so the solar system is entirely in the $xy$-plane, the $z$-coordinate of all planets will be zero, and we can specify the position using only the $(x, y)$ coordinates. Note that the coordinates themselves carry units (in this case, AU) and can be positive or negative; these numbers represent the *positions* along the respective coordinate axes, not their *distances* from the origin.

---

### 1.6.1. Activity: Planetary Coordinates

**a.**  Fig. 1.3 shows the inner portion of our solar system on a set of coordinate axes. The sun sits at the origin, and each tick mark represents one-tenth of an AU. For convenience, the figure also contains dotted circles that are located every half AU. Specify the $x$ and $y$ coordinates of all four planets according to this coordinate system. Give your answer as an $(x, y)$ pair (don't forget units), and try to be reasonably accurate (a ruler might help).

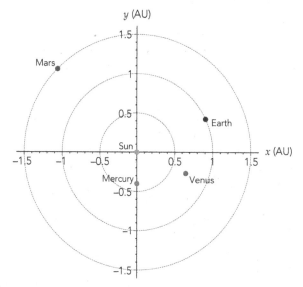

**Fig. 1.3.**  The inner planets of our solar system are shown using a coordinate system with the sun at the origin.

**b.** Notice that Mars lies on a circle with a radius of approximately 1.5 AU from the sun. In general, can you determine a formula that gives the distance of an object from the origin in terms of its Cartesian coordinates? **Hint**: Form a right triangle. Check to see that your distance formula works (or at least gets close) using your coordinates of Mars.

### A New Coordinate System

As mentioned, the choice of a coordinate system, including the location of the origin, is arbitrary; we are free to choose any coordinate system we want. In Fig. 1.3, we chose the origin to be at the location of the Sun and for the coordinate system to be oriented in such a way that the planets are all in the $xy$-plane. If we wanted, we could further orient the coordinate system so that Earth lies along one of the axes, as shown in Fig. 1.4.

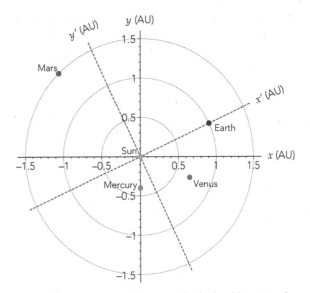

**Fig. 1.4.** A new coordinate system (shown with dashed lines) is chosen so that Earth lies along the $x'$-axis.

We have rotated the original coordinate system (represented by the $x$ and $y$ axes) into a new coordinate system (represented by the $x'$ and $y'$ axes) so that the Earth lies along the $x'$-axis. This "primed" coordinate system is just as valid as the original coordinate system, and may, in fact, simplify the situation.[10] For example, in the primed coordinate system the coordinates of Earth are given by (1 AU, 0 AU).

---

[10] The "prime" here should not be confused with the prime notation often used in calculus to denote a derivative. In our situation, the prime simply denotes a new $x$ and $y$ axes.

### 1.6.2. Activity: New Planetary Coordinates

Determine the coordinates of Mars (again, using a ruler) using the primed coordinate system and calculate its distance from the sun using your distance formula. If you're careful, you should find that your distance is close to what you found previously.

## 1.7   INTRODUCTION TO VECTORS

It turns out that many quantities in physics have a direction associated with them. A simple example is the direction your car is moving as you drive down the road. Another example, as just considered, is the positions of the planets in the solar system. While the *distance* of each planet from the sun can be specified without reference to a direction, the *position* of a planet in its orbit requires a magnitude *and* a direction.

Mathematically, these quantities are described using something called a *vector*. A vector can be thought of as an arrow of a certain length pointing in a particular direction.[11] A vector has a head (tip of the arrow) and a tail (end without the arrow). We will designate vectors using an arrow over the top of a letter/symbol. For example, we might represent a vector as $\vec{A}$ or $\vec{b}$. As mentioned, a vector has both a magnitude, represented by its length, and a direction, often represented by an angle. The magnitude of a vector $\vec{A}$ is always positive and is written $|\vec{A}| = A$ (the letter without the arrow); to determine its direction, we must specify the angle with respect to an axis in our coordinate system.

### Vector Components

It is important to be able to specify a vector using its magnitude and direction, but also in terms of its *components*. In fact, being able to move back-and-forth between these two descriptions is an important skill that you should master because this procedure comes up over and over in this course. To understand vector components, we consider a two-dimensional vector and consider its *projection* along the $x$ and $y$ axes (see Fig. 1.5). For a given vector $\vec{A}$, the projection along the $x$-axis is known as the *x-component* and is labeled $A_x$, while the projection along the $y$-axis is known as the *y-component* and is labeled $A_y$.

---

[11] Technically, vectors are defined according to their transformation properties; simply having a magnitude and direction is not sufficient. However, for most purposes it is fine to think of a vector as an arrow pointing in a particular direction.

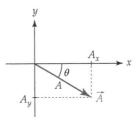

Fig. 1.5. A vector $\vec{A}$ is drawn showing its magnitude and components.

Some important comments about vector components:

- Components can be positive or negative, depending on whether the vector points along the positive or negative direction of a particular axis. In Fig. 1.5, the $x$-component of $\vec{A}$ is positive, while the $y$-component is negative.

- If the magnitude of $\vec{A}$ and the angle $\theta$ are known, the magnitudes of the components can be calculated using trigonometry. Whether these components are positive or negative is determined by whether the projection is along the positive or negative axis. In Fig. 1.5, $|A_x| = A\cos\theta$ and $|A_y| = A\sin\theta$ (recall that the magnitude of the vector, $A = |\vec{A}|$, is always positive). The signs of these components are then determined by the orientation of $\vec{A}$ to the coordinate system. For the situation given in Fig. 1.5, we see that $A_x$ is positive while $A_y$ is negative, giving $A_x = A\cos\theta$ and $A_y = -A\sin\theta$. Be aware that depending on how the angle is chosen, it could be that $|A_x| = A\sin\theta$ and $|A_y| = A\cos\theta$.

- If the angle $\theta$ is measured counter-clockwise with respect to the positive $x$-axis, then the components are *always* given by $A_x = A\cos\theta$ and $A_y = A\sin\theta$, with the signs automatically determined by the trigonometric functions. (While this is a guaranteed method for getting the correct components, most students prefer working with angles between 0 and 90°, and then determining the signs of the components by the orientation of the vector.)

- Since the $x$ and $y$ axes are perpendicular, $|A_x|$ and $|A_y|$ form the legs of a right triangle. From the Pythagorean theorem, the magnitude of the vector is then given in terms of its components as $A = \sqrt{A_x^2 + A_y^2}$.

### 1.7.1. Activity: Vectors and Components

Finding the components of a vector comes up quite frequently in physics, so it is worth some practice. For each of the examples below, draw the vector on the provided coordinate system (with units in meters), and then find the components of the vector given the vector magnitude and direction (or vice versa).

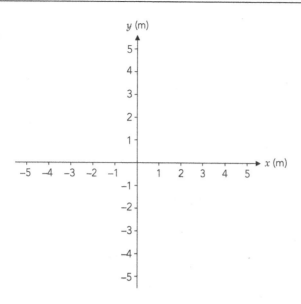

**a.** Vector $\vec{A}$ has length 3.0 m and points 70° above the positive $x$-axis.

**b.** Vector $\vec{B}$ has length 4.0 m and points 45° below the negative $x$-axis (45° below the negative $x$-axis is equivalent to 225° with respect to the positive $x$-axis).

**c.** Vector $\vec{C}$ has components (2.0 m, −4.0 m).

**d.** Vector $\vec{D}$ has components (−5.0 m, 3.0 m).

---

It is worth emphasizing once again that the ability to determine components from vectors specified by magnitude and direction (and vice-versa) is *extremely important*, so you should make sure you are comfortable with this activity before moving on. Mastering this skill is one of the biggest hurdles students face when learning physics.

## Vector Addition

Vectors can be added together by placing the vectors "head-to-tail," with the resultant vector going from the tail of the first vector to the head of the second (see left side of Fig. 1.6, where $\vec{A} + \vec{B} = \vec{C}$). Alternatively, one can use the "parallelogram method" (see right side of Fig. 1.6). In both cases, the resulting vector $\vec{C}$ is the same. Vector subtraction works similarly, although one typically thinks about adding the negative of a vector instead: $\vec{C} = \vec{A} - \vec{B} = \vec{A} + (-\vec{B})$, where $-\vec{B}$ is a vector with the same magnitude (length) as $\vec{B}$ but pointing in the opposite direction (all components switch signs).

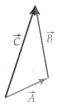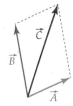

**Fig. 1.6.** Vector addition using either the head-to-tail method (left) or the parallelogram method (right). Both methods result in the same final vector $\vec{C} = \vec{A} + \vec{B}$.

### 1.7.2. Activity: Some Vector Math

**a.** Consider two vectors $\vec{A}$ and $\vec{B}$ with components $(A_x, A_y) = (1.0\,\text{m}, 4.0\,\text{m})$ and $(B_x, B_y) = (2.0\,\text{m}, -3.0\,\text{m})$. Using the coordinate system below, draw vectors $\vec{A}$ and $\vec{B}$ with the vectors located at the origin (the location of a vector is specified by the position of its *tail*).

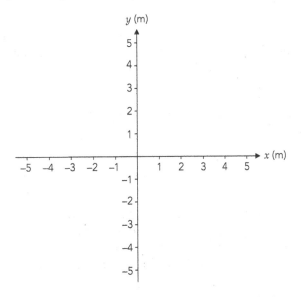

**b.** Use either the "head-to-tail" or "parallelogram" method to add these vectors graphically on your coordinate axes. Label the sum as vector $\vec{C} = \vec{A} + \vec{B}$.

**c.** Graphically determine the *components* of vector $\vec{C}$ and describe how these components are related to the components of $\vec{A}$ and $\vec{B}$ (you should find a very simple relationship).

Notice that when you add vectors, either head-to-tail or using the parallelogram method, you are effectively moving one of the vectors so that its tail no longer lies at the origin but instead is placed at the head of the other vector. The reason we can do this is because a vector is defined by its direction and magnitude (or its components); as long as these quantities remain the same, the vector will be the same no matter where it is located. It is often *useful* to position a vector at the origin, but it is not a requirement.

**Unit Vectors**

Another way of writing a vector $\vec{A}$ with components $A_x$ and $A_y$ is to use something called *unit vectors*: $\vec{A} = A_x\hat{x} + A_y\hat{y}$. In this expression, the symbols $\hat{x}$ and $\hat{y}$ (pronounced "x-hat" and "y-hat") represent *unit vectors* in the x and y directions, respectively.[12] The term unit vector (and the hat) is used to denote any vector that has a magnitude of one (they have "unit" magnitude, hence the term unit vector). Written this way, the quantity $A_x\hat{x}$ represents a vector of length $|A_x|$ pointing in either the positive (if $A_x$ is positive) or negative (if $A_x$ is negative) x-direction (similarly for $A_y\hat{y}$).

Note that the vector expression $\vec{A} = A_x\hat{x} + A_y\hat{y}$ is consistent with how we defined vector addition above: the vector $\vec{A}$ can be considered the sum of the two (perpendicular) *component vectors*, $A_x\hat{x}$ and $A_y\hat{y}$, added head-to-tail. Component vectors turn out to be pretty useful, so make sure you understand how they differ from a vector's *components*.

**1.7.3. Activity: Some More Vector Math**

Consider a coordinate system with the x-axis pointing east and the y-axis pointing north. You start at the origin of the coordinate system and ride your bike east for 3 kilometers. You then turn north and ride for two more kilometers.

**a.** Using the axes below, draw three vectors in the coordinate system: one vector starting at the origin and representing your initial ride east, one vector starting where you ended your ride east and representing your ride north, and a final vector representing the sum of these two vectors.

---

[12] There are a variety of different notations that are used for Cartesian unit vectors. In this text, we will use $\hat{x}, \hat{y}, \hat{z}$, but $\hat{i}, \hat{j}, \hat{k}$ and $\hat{e}_x, \hat{e}_y, \hat{e}_z$ are also commonly used to represent Cartesian unit vectors.

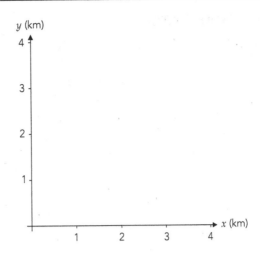

**b.** Let's call the third vector (the sum) $\vec{r}$. Write $\vec{r}$ in component form using unit vectors (don't forget to include units).

**c.** Vector $\vec{r}$ represents your *net displacement* (from the origin) after both portions of your ride. In other words, it shows where you ended up relative to where you started, independent of how you got there. At the end of your ride, what is your *distance* from the origin? How does this compare to how far you actually rode? Be sure to show your work and explain your answer.

**d.** What angle does vector $\vec{r}$ make with respect to the positive $x$-axis? Again, be sure to show your work.

---

The vector $\vec{r}$ represents the net displacement from the starting position. Because our initial location was the origin, the vector $\vec{r}$ here corresponds to what we call the *position vector*.

## 1.8   THE POSITION VECTOR

One of the most important vectors in the study of mechanics is the position vector, typically denoted by $\vec{r}$. This vector points from the origin to the location of an object and represents the position of the object with respect to the origin (see Fig. 1.7).

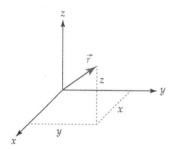

**Fig. 1.7.** The position vector in three dimensions points from the origin to the location of interest.

For an object located at an arbitrary position, we use the variables $x$, $y$, and $z$ to represent the coordinates of the object. The position vector is then written as

$$\vec{r} = x\hat{x} + y\hat{y} + z\hat{z} \qquad (1.3)$$

For example, suppose an object is located at the coordinates $(x, y, z) = (1.6$ m, $1.3$ m, $2.2$ m$)$. The position vector is then $\vec{r} = 1.6$ m $\hat{x} + 1.3$ m $\hat{y} + 2.2$ m $\hat{z}$. Note that with this definition, the components of $\vec{r}$ are the same as the coordinates of the object's location:[13] $(r_x, r_y, r_z) = (x, y, z) = (1.6$ m, $1.3$ m, $2.2$ m$)$.

---

### 1.8.1.  Activity: Even More Vector Math

**a.**  Look back at Activity 1.6.1 part (a), where you found the $x$ and $y$ coordinates of the four planets. Write down the position vector (in three dimensions) for each of the four planets using unit vectors. (This is quick, but don't forget units!)

**b.**  Figure 1.7 shows the position vector for an object at an arbitrary location in three-dimensional space. Determine an expression that represents the *distance* of that object from the origin. Be sure to show your work.

---

[13] We note here a potential point of confusion: a triplet of numbers, such as $(a, b, c)$, represents a set of *coordinates* that specifies a particular point in space within a given coordinate system. However, this same notation is sometimes used to represent a *vector* (represented by an arrow), where $(a, b, c)$ are the *components* of the vector. Because of this potential confusion, one should always be clear whether the triplet specifies a set of coordinates or the components of a vector.

**Hint**: If you're confused about how to begin, start by finding the distance from the origin to the point $(x, y, 0)$.

c.  Now consider arbitrary position vectors for two different positions: $\vec{r}_1 = x_1\hat{x} + y_1\hat{y} + z_1\hat{z}$  and  $\vec{r}_2 = x_2\hat{x} + y_2\hat{y} + z_2\hat{z}$. Find an algebraic expression for the vector $\vec{r}_2 - \vec{r}_1$.

d.  Now find the *distance* between the two positions. Note that this is the *length* (or magnitude) of the vector $\vec{r}_2 - \vec{r}_1$. **Hint**: Using what you found in parts (b) and (c) will make this easy!

---

**Additional Vector Properties**

In the sections above, we added (and subtracted) vectors in a manner that hopefully seemed clear to you. But what about multiplication? Can two vectors be multiplied together? The answer is yes, but it's probably not obvious exactly how to do this. It turns out that there are two common ways of multiplying vectors;[14] however, we will hold off discussing vector multiplication until it becomes necessary.

---

[14] Interestingly, while there are two common ways of multiplying vectors, vector division is not defined at all! In fact, it turns out that there are more than two ways for multiplying vectors, but in this course we will only make use of two.

### 1.9  PROBLEM SOLVING

In this unit, we have discussed topics including unit conversions, scientific notation, significant figures, the orbital period of satellites and planets, coordinate systems, and vectors. It is now time to put everything together and solve some problems. The following questions should be answered as completely and accurately as possible. Please be sure to show your work, such as drawing any relevant diagrams, writing down the starting equation(s), and briefly explaining your steps.

nasaimages / 123 RF

#### 1.9.1. Activity: Jupiter

a. Jupiter is the largest planet in our solar system with a (mean) radius of $7.13 \times 10^7$ m and a mass of $1.90 \times 10^{27}$ kg; it orbits the sun at a distance of 5.20 AU. Using only information contained in this unit, determine the orbital period of Jupiter in (Earth) years.

b. Because we are free to choose any coordinate system we want, at any moment in time we can choose our axes such that the sun is at the origin and Earth lies along the positive $x$-axis. Suppose at this instant of time, Jupiter makes an angle of 120° (counter-clockwise) with respect to the positive $x$-axis. Determine the distance between Jupiter and Earth (in AU) at this time. **Hint**: A diagram might be helpful.

# UNIT 2: MEASUREMENT AND UNCERTAINTY

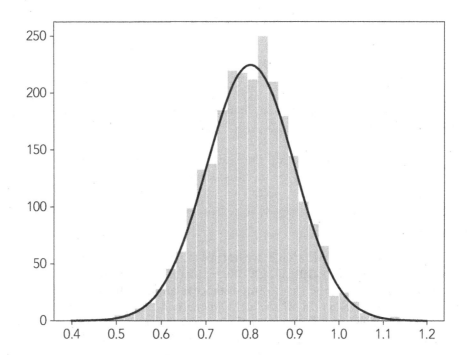

*Physics is an experimental science, and all the theories developed by physicists are ultimately based on observations and measurements. It is therefore possible for two investigators to measure the same physical quantity and come up with two different values. Although neither investigator has made any measurement errors, they could still decide that their measurements are effectively the same. After completing this unit, you should be able to explain why scientists making different measurements can agree that they have found the same value.*

# UNIT 2: MEASUREMENT AND UNCERTAINTY

## OBJECTIVES

1. To learn how to estimate, quantify, and minimize sources of random uncertainty so that the precision of measurements can be improved.

2. To learn about systematic error in measurements and how it differs from random uncertainty.

3. To explore the mathematical implications of the standard deviation and the standard deviation of the mean associated with a set of measurements.

## 2.1  OVERVIEW

In our discussion of the solar system in Unit 1, we learned that the distance between the center of Earth and the center of the Sun is $1.50 \times 10^{11}$ m. Of course, this is not the *exact* distance between Earth and the Sun. After all, measuring the Earth–Sun distance is not a simple task; we can't just stretch a tape measure between the two and read off a value! One method for measuring this distance is to send out radio waves and detect the signals that are reflected off other planets. By making use of relations between the interplanetary distances, one can then *calculate* the Earth–Sun distance. In this way, the Earth–Sun distance is measured *indirectly* and, as you might guess, it is impossible to determine the exact distance using such a method.[1] When defining the Astronomical Unit (AU), scientists use the best measurements available for the average distance between the Sun and Earth.

There are a few quantities that can be determined, at least in principle, with complete certainty. For example, if discrete items are counted, the degree of *precision* can be infinite: everyone can agree, for example, if there are two (or three) cats sitting on a couch. In addition, certain mathematical quantities, such as $\pi$ (the ratio of the circumference to the diameter of a circle), can be determined to any desired precision. Such infinite precision is possible because a circle is an abstract, idealized mathematical object.

However, the *measurements* you make in the study of the physical world, such as the length of an object, the temperature of a liquid, or the time it takes a ball to fall to the ground, will only be estimates of the true values we are trying to find. There are inevitable *uncertainties* associated with such measurements, as there are no instruments capable of infinite precision when the quantities being measured can take on any value. The result is that all measurements have some amount of uncertainty. Therefore, an important aspect of making measurements of any kind is to include an estimate of the uncertainty

---

[1] In addition, the Earth–Sun distance varies through the year as Earth orbits the Sun. So even if one could measure an exact distance, it would change throughout the year.

associated with the measurement. For a direct measurement (e.g., the length of a small object using a ruler) this is usually not too difficult, but for indirect measurements the procedure can be quite complicated. Although we will not spend time in this course making detailed uncertainty calculations, it is important to understand the basic ideas involved in estimating, and eventually calculating, such uncertainties. Knowing the uncertainty is critical when deciding whether two measurements "agree" with each other, or whether a measurement is compatible with a particular theory.[2]

---

[2] Although there is an implied uncertainty for any measured value, it is often the case that the actual uncertainty is larger than what the implied uncertainty would be. For this reason, it is a good idea to include an uncertainty estimate with any measurement.

## MEASUREMENTS AND UNCERTAINTY

### 2.2 MEASURING LENGTHS DIRECTLY

Suppose we are interested in finding the length of a particular object. For example, consider using a ruler to measure the length of a small item such as a key. The uncertainty associated with such a measurement will depend on the type of ruler you use: the finer the scale on the ruler, the smaller the uncertainty. For the activities in this section, you will need the following equipment:

- 1 small object (key, cell phone, etc.)

#### 2.2.1. Activity: Length Measurements

**a.** Using a small object such as a key or a cell phone, measure its length (or width) using the ruler shown in Fig. 2.1 (and nothing else). What value would you report using this scale? What is your estimated value of uncertainty? Give your answer as a length ± the uncertainty, being sure to include units. **Hint**: Remember our earlier discussion of scientific notation and implied uncertainty.

Fig. 2.1. A ruler with a very coarse scale.

**b.** Now repeat the measurement and uncertainty estimate using the ruler shown in Fig. 2.2.

Fig. 2.2. A ruler with a finer scale.

**c.** Finally, repeat the measurement and uncertainty estimate using the ruler shown in Fig. 2.3.

Fig. 2.3. A ruler with a much finer scale.

In the preceding activity it should be clear that the estimated uncertainty depends on the measuring tool. For the first ruler, the precision of your measurement was limited by the fact that the ruler only has lines marked every centimeter. If the end of the object was located between two ruled lines, you would have to make an estimate for the actual size of the object. For example, if the end was just a little past the 2 cm mark, you could safely say that the object was closer to 2 cm than 3 cm. However, you might not be able to do much better than that. If this is the case, you might report a length of 2 cm, with an uncertainty of half a centimeter: $2 \pm 0.5$ cm.

On the other hand, maybe you feel comfortable saying it was definitely between 2.0 cm and 2.4 cm. In this case, you might choose to report a value of $2.2 \pm 0.2$ cm, where 2.2 cm is in the middle of the region. In reporting your measurement, you should only include digits that are meaningful (significant). You can use the estimated uncertainty to help decide this. For example, if your uncertainty was $\pm 0.2$ cm, you should not report any numbers past the tenths digit. Why? Because if the tenths digit is uncertain, the hundredths digit would be essentially meaningless.

Hopefully, it's clear that a ruler with more divisions allows for a more precise length estimate. However, it's impossible to make even a simple distance measurement without some uncertainty. Although we can increase the precision of our measurement by using a ruler that has more divisions, there is always a limit on how many divisions can be placed on the ruler (and similar limits exist for all measurement tools). This measurement uncertainty is an example of an *inherent* uncertainty, and no matter how hard you try such uncertainties can never be completely eliminated.

## 2.3  UNCERTAINTY AND ERRORS

In common terminology there are three kinds of uncertainties that arise in measurements: (1) *inherent uncertainties* (like we just discussed), (2) *systematic errors* due to measurement or equipment problems, and (3) *mistakes* (or "human errors"). For the activities in this section, you will need the following equipment:

- 1 meterstick (for every student in the class, if possible)

### 2.3.1.  Activity: "Human Error"

Give an example of how a person could make a mistake, or a "human error," while making a length measurement.

Clearly it is possible for anyone to make a mistake while making a measurement, performing a calculation, or even when writing down the result. However, such mistakes can, at least in principle, always be eliminated by being very careful, checking our work, having someone else check our work, etc. Therefore, throughout this book we will assume there are no "human errors" and instead focus on inherent uncertainties and systematic errors.

### Inherent Uncertainties

The limitations of the rulers in Section 2.2 lead to inherent uncertainty; one can only be so precise with the measuring tool available, so there is always some uncertainty *inherent* in the measurement process. Such inherent uncertainties do not result from mistakes or errors. Instead, they are attributed (at least in part) to the impossibility of building measuring equipment that is precise to an infinite number of significant figures. The ruler provides us with an example of this. It can be made better and better, but it always has an ultimate limit of precision.

There are also examples of inherent uncertainties that are not related to the measuring device. For instance, if you measure the width of a door, you might find that you get slightly different values depending on where the width is measured. Clearly, a door in the real world is not going to be a perfect rectangle with opposite sides being exactly parallel. We know that there will be imperfections in the door, and these imperfections lead to an inherent uncertainty in quantities such as its width. In some sense, the door doesn't really have a perfectly well-defined width. Finally, inherent uncertainty can also be part of the process being studied, as we will discuss later.

### Systematic Errors

Systematic errors result when some type of error occurs over and over again in a *systematic* way. For example, suppose you are making a distance measurement and use a ruler that was poorly calibrated. In this case, a careful reading of 5.0 cm on the ruler might actually correspond to a true measurement of 4.9 cm. No matter how careful you are, your result will still be "wrong." Systematic errors can also result from an uncorrected "human error." A common example is when the zero on a ruler does not occur at the end of the ruler, but slightly inside the edge (see Fig. 2.4). If you line up the end of this ruler with the side of the object (instead of at the zero mark), each measurement you make will be a little smaller than the actual length of the object.[3]

Fig. 2.4. The end of this ruler does not correspond to the beginning of the measurement scale, so any measurement made by lining up the edge of the ruler to the object will be systematically smaller than it should be.

Fortunately, a systematic error can usually be corrected for, assuming you are aware of it. For example, if you were using the end of the ruler in Fig. 2.4 to make measurements and then realized that the zero line does not occur at the end of the ruler, you could measure the distance between the end of the ruler and the zero line and add this value to your previous measurements. Of course, some systematic errors can be extremely difficult to locate, and in most cases you won't even be aware that a systematic error is occurring. This is especially true when using a measurement device more complicated than a ruler. If you don't know the details of what is happening inside the device, it is very challenging to determine whether the equipment is giving "correct" readings.

---

[3] Even though such an error is, in some sense, a "human error," we will still avoid this term for these situations, referring to it instead as a systematic error. Incidentally, the zero mark on a ruler is not usually aligned with the edge of the ruler because the edge can get worn down, which will lead to incorrect measurements.

### 2.3.2. Activity: Measuring the Size of the Room

**a.** The goal of this activity is to measure the width (or length) of your classroom in under two minutes using a single meter stick. After recording your value, repeat the entire process so you have two *independent* measurements. Each student should make their own measurements, and don't worry if they don't agree. Write down your results below and on the board at the front of the room. What *uncertainty* would you assign to your width measurement? Briefly explain how you came up with this uncertainty.

**b.** Now consider the entire set of measurements performed by the class (with perhaps 40–50 data points). Is it possible to determine whether there are any *mistakes* in this set of measurements? How would you know? If you were to discover a mistake, what would you do? Do there appear to be any mistakes in the class data?

**c.** Still considering the entire set of measurements, is it possible to determine if there are any *systematic errors* in the measurements? If you do notice what appears to be a systematic error, how might you check and correct for it? Do there appear to be any systematic errors in the class data?

**d.** Using the class data, how might you determine the *best estimate* of the room width? Briefly explain your process and carry out the procedure.

### The Histogram

When we have a set of measurements of the same quantity, it is often useful to plot what is known as a *histogram*, or a *frequency distribution* for the set of measurements. The histogram shows how often similar values are measured.

The horizontal axis of the histogram indicates the quantity being measured (e.g., the width of the room in meters), while the vertical axis gives the number of times each measurement occurs.

Because it is rarely the case that the exact same value is measured multiple times, the numbers on the horizontal axis are typically "binned" into ranges so that all values within a certain range are counted together. For example, suppose that each student in the class measured the length of an object using a coarse ruler as in Activity 2.2.1 (a). We might choose to sort the measurements into "bins" of 1 cm width for plotting the histogram. Thus, the bin labeled "3 cm" would include all measurements from 2.5 cm to 3.5 cm, while the bin labeled "4 cm" would include all measurements from 3.5 cm to 4.5 cm, and so on. Figure 2.5 shows a hypothetical histogram with the results of such measurements, where four students in the class reported a value between 2.5 cm and 3.5 cm, two students reported a value between 3.5 cm and 4.5 cm, and four students reported a value between 4.5 cm and 5.5 cm.[4]

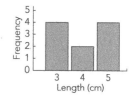

**Fig. 2.5.** A sample histogram showing a set of length measurements.

To see how this works in detail, let's make a histogram of the room width measurements carried out in the previous activity. Your instructor may provide specific directions on how to accomplish this.

---

### 2.3.3. Activity: The Frequency Distribution of a Set of Measurements

**a.** Collect the room width measurement results for the entire class into a spreadsheet program and create a histogram of the results, choosing appropriate bin ranges so that you end up with roughly 10 bins. Do you see any outliers in the data that might indicate either a mistake or a systematic error? If you find an *obvious* mistake, you can remove that data point from the set.[5] If you find what you believe to be a systematic error for some of the measurements, make sure you correct for it (if possible). Briefly explain any changes you make below.

**b.** Once you have removed any spurious data points and corrected for any systematic errors, print out a copy of your histogram below (or make a sketch) and comment on its shape. Do you think the shape tells you anything about the measurement process or your final answer? Explain briefly.

---

[4] Note that for demonstration purposes, we chose a set of measurements that has a larger amount of spread than would be typical!

[5] One must be very careful when removing a data point from a set of measurements. You need to be certain that it is an actual mistake; it is ***not*** okay to remove a data point simply because you don't like it!

c. For a set of $N$ data points, you probably know that the *average* (technically the *arithmetic mean*) is calculated by adding up the individual measurements and dividing by the number of measurements. Mathematically, this can be represented by the following expression

$$\langle x \rangle \equiv \frac{x_1 + x_2 + \cdots + x_N}{N} = \frac{1}{N}\sum_{i=1}^{N} x_i$$

In this equation, $\langle x \rangle$ represents the mean, the individual measurements are denoted as $x_i$, and the large "sigma" represents a summation (you *sum* up the values of $x_i$ as $i$ runs from 1 to $N$). Use the spreadsheet to calculate the average of the class measurements, making sure you have already removed any mistakes and corrected any systematic errors. (Most spreadsheet programs have a built-in function for calculating the average, typically called "AVERAGE.") Mark the position of the average on your histogram above by drawing a vertical line. Where does the average fall within the distribution? Is the average a better estimate for the width of the room than any individual measurement? Briefly explain why or why not.

d. You should find that the data is "scattered" about the average. Do you think this scatter is related to measurement uncertainty in some way? Briefly explain why or why not.

As you may already know, the average of a set of measurements does not tell the entire story. For example, if everyone took their time and was extremely careful when measuring the width of the room, then presumably all the measurements would have nearly the same value. In such a situation, there would be very little scatter in the data, so the average would not differ much from each of the individual measurements—there is little uncertainty in the individual measurements. However, if each of the measurements varies from the others by a large amount, this is a sure sign that there is more uncertainty in the individual measurements. In this case, taking an average would provide a much better estimate for the "true" value we are trying to measure. The reason for this is because some measurements are larger than the average while others are smaller, so taking an average acts to "cancel out" values that are too large with those that are too small, resulting in a more trustworthy value than any single measurement. To make these ideas more concrete, the next activity attempts to quantify the amount of scatter (or "width," or "spread") in a data set.

## 2.4   STATISTICS AND QUANTIFYING UNCERTAINTY

At first glance, it might not be obvious how to go about quantifying a "width" for the histogram. After all, the plot probably falls off to either side gradually so that there is no obvious "edge" to the data set. However, it should be clear that the width is related to how "far away" the points are from the average of the data set.

MarekPhotoDesign.com/
Adobe Stock

### 2.4.1.  Activity: The Standard Deviation

**a.**  One possibility for quantifying the spread in the data is to calculate the difference between each data point and the mean, and then averaging the results. That is, you take each measurement and subtract the mean to determine how "far" this data point is from the average of the set. You then take the average of these differences (add them up and divide by $N$). Mathematically, this process can be written as

$$\frac{1}{N}\sum_{i=1}^{N}(x_i - \langle x \rangle)$$

Use your spreadsheet to calculate this quantity for your data set (your instructor may have some tips on how to accomplish this task). Does the final result seem to be a good measure for the width of your data set based on your histogram? Briefly explain.

**b.**  While the idea presented above may seem quite reasonable, your answer may be surprising. Explain why this procedure does ***not*** provide a good measure for the width of the distribution. **Hint**: The summation above can be broken up into two separate sums; what is the value of the first sum?

The problem with calculating the difference between each data point and the mean is that you get both positive and negative values, and these values tend to cancel each other out when taking the average. In fact, it's not too difficult to show that the summation given in part (a) is identically zero! What we really need to do is find the average "distance" between the data points and the calculated mean. The key point is that "distance" is always positive. One way to get such a positive measure is to average the differences *squared*, which guarantees all values will be positive. Such a procedure avoids the inadvertent cancellation you get from positive and negative values when averaging just the differences. This particular measure is called the *variance* and is expressed mathematically as:[6]

$$\mathrm{Var} \equiv \frac{1}{N-1} \sum_{i=1}^{N} (x_i - \langle x \rangle)^2$$

Of course, the units of the variance are not what we want for a width of the spread in our measurements: for a length measurement, the units of the variance will be length squared. But taking the square root of the variance will give a measure for the width of the distribution known as the *standard deviation*. The standard deviation is represented by the lowercase Greek letter sigma σ (often labeled $\sigma_{sd}$) and given mathematically by

$$\sigma_{sd} \equiv \sqrt{\frac{1}{N-1} \sum_{i=1}^{N} (x_i - \langle x \rangle)^2} \qquad (2.1)$$

  **c.** Use your spreadsheet program to calculate the standard deviation of the set of classroom length measurements and record its value below (including units). Again, most spreadsheet programs have a built-in function for this calculation (e.g., "STDEV"). Draw two vertical lines on your histogram plot that are separated by one standard deviation to either side of the average (the separation between the two lines will be twice the standard deviation). How does the standard deviation compare to the width of your distribution? Does the standard deviation appear to be a good indicator for the uncertainty in these measurements?

As this activity demonstrates, the standard deviation $\sigma_{sd}$ provides a reasonable measure for the width of a distribution. Although the distribution doesn't have sharp edges, the standard deviation provides a rough estimate for the spread in the measurements. As we will see, the standard deviation also provides some "level of confidence" about what you will find when making additional measurements.

### 2.4.2. Activity: Improving Our Measurement

  **a.** Suppose you make 10 measurements of the width of the room and calculate the average. Now consider making 100 measurements and calculating the average. Would one of these two averages provide a better estimate for the true value you are trying to measure? Why or why not?

---

[6] For technical reasons relating to the fact that our sample mean is based on a finite number of measurements, it turns out to be better to divide by $N-1$ instead of $N$ when calculating the variance. Note that for large values of $N$ (many measurements) this change has very little effect on the final result.

(By "better estimate" here, we mean a value that is more trustworthy in that it has a *smaller uncertainty*.)

**b.** Two histograms of hypothetical data for a room measurement experiment are shown below, one consisting of 10 measurements, and one for 100 measurements. Do the standard deviations for these two data sets appear to be the same or different? **Note:** You are *not* supposed to actually calculate the standard deviations here, you're just supposed to "eyeball" it.

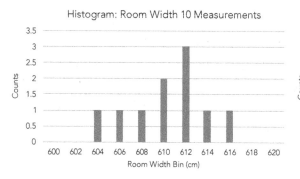
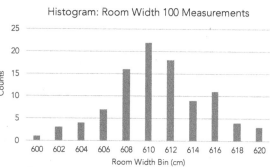

**c.** Based on the data in part (b), do you think the standard deviation is a good measure for the uncertainty of the *average*? Why or why not?

You may not be completely confident in your answer to the last question in the previous activity. That's okay, we will return to this question later in the unit.

# NORMAL DISTRIBUTIONS AND NUCLEAR DECAY

## 2.5   NATURAL RADIOACTIVITY AND STATISTICS

In the last activity, we made a series of measurements for the length of the room and compiled the data. After eliminating mistakes and correcting for any apparent systematic errors, we ended up with a data set that, when viewed as a histogram, looks a little like a bell-shaped curve. Moreover, we saw that the standard deviation provides a measure of the spread in the data set and may be a reasonable estimate for the measurement uncertainty.

The most common bell-shaped distribution is known as a *Gaussian* distribution, also called a *normal* distribution.[7] Gaussian distributions typically govern processes that are *random* (or *stochastic*) in some way. It turns out that many natural phenomena follow a Gaussian distribution, and the measurement process itself, which is governed by random uncertainties, also results in a Gaussian distribution.[8] Because Gaussian distributions are so common in science, it is worth investigating them a little more closely. To do so, we will need a quick and easy method of obtaining many data sets that obey Gaussian statistics. One convenient way of producing such data sets is to measure radioactive decay of nuclei.[9]

To begin, we need to briefly discuss radioactivity. Everything in the world is made up from *elements* that appear in the periodic table. Each element is distinguished by the number of protons in the nucleus. But for a given number of protons, the element can have different numbers of neutrons in the nucleus, which are referred to as different *isotopes* of that element. While some of these nuclei are stable, it turns out that many naturally occurring isotopes have nuclei that are *unstable*. An unstable nucleus can lower its energy by reconfiguring itself into something a little different; literally, one material is transformed into something else. This process of nuclear reconfiguration is what we call *radioactivity*, and it can be understood as follows: every once in a while, a particular nucleus in a collection of radioactive (unstable) atoms *decays* by ejecting either a gamma ray, a beta particle, or an alpha particle. Radioactivity is a statistical process in which a series of random disturbances in a nucleus lead to its decay.

Heavy elements such as uranium and thorium occur naturally in rocks and soil, and even a few tiny grains of such an element can contain hundreds of billions of radioactive nuclei. The subatomic particles ejected from a sample of radioactive matter can be counted using a *Geiger tube*, which forms the heart of what we will refer to as a *radiation sensor*.

---

[7] The origin of the term "normal" distribution is not terribly clear, but it appears to be due to Francis Galton who, in his 1889 book *Natural Inheritance*, began systematically using the term to describe a sense of conforming to a norm.

[8] Although Gaussian distributions are extremely common, there are many other types of probability distributions that govern particular phenomena, so there may be times when assuming a Gaussian distribution is not appropriate. Even still, it is often the case that non-Gaussian distributions are well approximated as Gaussian.

[9] Technically speaking, the process of nuclear decay follows what's called a *Poissonian* distribution, but this is nearly indistinguishable from a Gaussian distribution when the average is approximately 20 or higher.

### Nuclear Counting

Each radioactive material decays according to the details of its nucleus. Typically, a radioactive material is specified by its *half-life*, which is the length of time it takes for half of the material to decay into something else. As you may already know, there is a vast range of half-lives for different radioactive materials. Some half-lives are very short (a few milliseconds), while others are extremely long (billions of years). When the half-life is short, you can actually observe the material disappearing because the count rate detected by the radiation sensor will decrease with time. However, if the half-life is long, then the count rate will not change appreciably over short periods of time. For example, if the half-life of a material is 30 years, then it would take 30 years for the count rate to be cut in half. Thus, over the period of a few hours (or even a few days), the average count rate would not appear to change at all. For the activities that follow we will be using materials with relatively long half-lives so the average count rates can be assumed to be constant.

In the next activity, we will count the number of decay particles coming into a Geiger tube during a fixed time interval. Then we'll do it again (and again, and again, and again). You can use this set of measurements to calculate the average count rate, the standard deviation, and produce a histogram. The main point of these experiments is to gain a more quantitative understanding of the Gaussian distribution and whether the standard deviation is an appropriate way to quantify the uncertainty for a particular measurement.

For this experiment, your group will need a radioactive source and a radiation sensor interfaced to a computer to automatically count the particles coming into the sensor. This setup allows us to take large quantities of data easily and quickly so that we can focus our attention on the distribution itself.

### Measuring Counts per Time Interval

In the exploration of the statistics of radioactivity, your group will need the following equipment:

- 1 radiation sensor with attached computer interface (Fig. 2.6)
- 1 low-level radioactive source

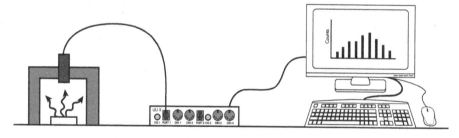

**Fig. 2.6.** A radiation monitoring system interfaced to a computer.

Your instructor may provide details on how to use the system with your computer. Set up the monitoring system and adjust the distance between the radioactive source and the radiation sensor (as well as the time interval) until you are getting approximately 30 counts/interval. You can select whatever time interval you need (0.1 seconds, 0.5 seconds, 1 second, etc.), though a shorter time interval will result in shorter experiments. See what happens when you move the source farther away from the Geiger tube in the radiation sensor. Once the distance from source to detector is adjusted to give an appropriate number of counts in the chosen time interval, *the source and the radiation sensor should not be disturbed*. **This is important**!

---

### 2.5.1. Activity: Is There Random Variation in Nuclear Counting Data?

**a.** Begin by adjusting the *total* collection time to produce an experiment with 100 separate data points, each of which gives the number of counts in your time interval. Comment on what you observe. For example, do the counts appear to be getting systematically smaller or larger over time, or are they roughly constant with some variation?

**b.** Plot a histogram of your results, either by using the radiation counting software or by transferring the data to a spreadsheet. Determine the average counts/interval and the standard deviation for your data set and record these values below.

**c.** Next, determine the percentage of data points that lie within plus or minus one standard deviation of the average. You can do this using the histogram or by sorting your data. In either case, briefly explain how you determined this number. Once you have this information, write it on the board at the front of the room.

**d.** By looking at the results from all groups, we want to come up with a "practical meaning" for the standard deviation $\sigma_{sd}$. In other words, if you were to make *one additional measurement*, what is the (approximate) likelihood that this value would fall within one standard deviation from the average? What does this tell you about the standard deviation in regard to the uncertainty of a single measurement?

**e.** Finally, take advantage of the power of the computer-based radiation monitoring system and extend the experiment time so that you obtain several hundred (or several thousand) data points. Does the histogram look like the classic bell-shaped curve? Print out (or sketch) a copy of your histogram in the space below. Mark the average along with the range $\langle x \rangle \pm \sigma_{sd}$. Does it seem like the same percentage of data points lies within one standard deviation of the average as measured in part (c)? (You don't actually need to count these, just estimate it.)

For a true Gaussian distribution, the percentage of data points that fall within one standard deviation of the mean will be 68.3%. Similarly, 95.4% of the data will fall within two standard deviations of the mean and 99.7% of the data will fall within three standard deviations of the mean. Thus, once we know the average and standard deviation, we have a good sense for what to expect when making another measurement. Namely, if we make one additional measurement, there will be a 68.3% chance that this value will fall within one standard deviation of the average, a 95.4% chance that it will fall within two standard deviations of the average, and a 99.7% chance that it will fall within three standard deviations of the average. This statement provides a very clear indication of what the standard deviation actually tells us (for a Gaussian distribution).

**Uncertainty in the Average**

In Activity 2.4.2, you were asked whether the average of a set of 10 measurements or a set of 100 measurements will provide a better estimate for the true value you are trying to measure. Most students have a strong sense that more measurements will result in a better average, and this is definitely true. In particular, the uncertainty in the 100-data-point *average* will be smaller than the uncertainty in the 10-data-point *average*. Yet it may not be clear if the standard deviations for the data sets in Activity 2.4.2 show this behavior or not, so let's investigate this question further.

### 2.5.2. Activity: Is the Standard Deviation a Measure of the Uncertainty in the Average?

a. Once again, set up your radiation sensor and radioactive source so that you are getting about 30 counts per interval. Once you have it set up, it is important not to disturb the source or the sensor for the entire activity! **Note**: If something gets bumped or moved you will need to start over, so be careful. Perform a series of experiments in which you measure 20 data points, 40 data points, 80 data points, 160 data points, and 320 data points (all without moving the source or the sensor). For each of these experiments, determine the average and the standard deviation and fill in the table below.

| # of Data Points | Average | Std. Dev. |
|:---:|:---:|:---:|
| 20 | | |
| 40 | | |
| 80 | | |
| 160 | | |
| 320 | | |

**b.** Which of these *averages* should give the best estimate for the true value; in other words, which average should have the smallest uncertainty? But what do you notice about the standard deviations? Do they *systematically* get smaller as you increase the number of data points? Explain briefly.

**c.** Explain how the results of this experiment clearly demonstrate that the standard deviation is *not* a good measure for the uncertainty of the average.

**d.** Although the standard deviation is not a good measure for the uncertainty in the average, it is still a good measure of something! In your own words, explain (once again) what the standard deviation actually tells us.

## 2.6 AVERAGES, UNCERTAINTY, AND CONFIDENCE INTERVALS

### Standard Deviation of the Mean

To get a good estimate of some quantity, you need to make several measurements and then take the average. This is the value we would report for our measurement. If we are also going to report an uncertainty for this measurement, we need to know how uncertain the *average* of our measurements is, since it is the average that we will write down as our best estimate. As we saw in the last activity, the standard deviation is *not* a good measure for the uncertainty in the average. Instead, the standard deviation is a measure of the uncertainty in a *single* measurement, and our average is made up from many such measurements. Therefore, the uncertainty in the average should be *smaller* than the standard deviation, and the more measurements we make, the smaller this uncertainty should be.

To determine a measure for the uncertainty in the average, let's think about the specific meaning of the standard deviation. The standard deviation (for a Gaussian distribution) tells us there is a 68.3% chance that one additional measurement will lie within one standard deviation of the average. Therefore, a good measure for the uncertainty in the *average* of a set of measurements would be the quantity that answers the following question: "If I repeat *the entire series of N measurements* and get a new average, what range will lead to there being a

68.3% chance that the new average lies within this range?" This range is determined by what's called the *standard deviation of the mean*, or SDM.[10]

How, exactly, do we go about finding the SDM? To answer this question, let's consider how we determined the standard deviation in the first place. We began with a series of measurements and plotted a histogram, and the width of the histogram led to our definition of the standard deviation. Thus, to determine the standard deviation of the mean we need a data set that consists of *averages*; we need to perform many separate experiments, each of which returns an average. We can then calculate the standard deviation for this set of averages, and this will provide a measure for the uncertainty in this data set. The following activity should clarify this idea.

### 2.6.1. Activity: Calculating the SDM the Hard Way

**a.** Using a radioactive source and a radiation sensor, set up an experiment that will measure the counts in a given time interval 50 times. As before, you should adjust the distance between the source and the sensor and the time interval until you are getting around 30 counts per interval. Once you have the source and sensor set up, *they must not be disturbed for the entire experiment*. Therefore, be sure to set up your experiment carefully so that it will not get bumped accidentally.

Once you are ready, run the experiment 10 separate times and determine both the mean and standard deviation for each experiment. Enter your results in the table below.

| Run # | Average (Mean) | Std. Dev. |
|-------|----------------|-----------|
| 1     |                |           |
| 2     |                |           |
| 3     |                |           |
| 4     |                |           |
| 5     |                |           |
| 6     |                |           |
| 7     |                |           |
| 8     |                |           |
| 9     |                |           |
| 10    |                |           |

**b.** Keep in mind that each of the averages in the above data table is a result of 50 separate measurements, and the standard deviations tell us the uncertainty associated with these *individual* measurements. Specifically, there is a 68.3% chance that any one of these individual measurements lies within one standard deviation of the mean, a range of $2\sigma_{sd}$. However, you should see that spread in the *averages* is confined to a much smaller range. Make an *estimate* of the range over which 68.3% of these

---

[10] The SDM is sometimes referred to as the *standard error*, but since the SDM is really a measure of uncertainty rather than error, we have chosen to avoid this term.

averages lie (no calculations needed here). In other words, what range contains about 7 of the 10 averages above?

c. To quantify the above observation, we want to calculate the standard deviation of this set of averages (the *standard deviation of the means*, or the SDM). The easiest way to do this is to copy the averages above into a spreadsheet program and use the built-in STDEV function. You should find that the SDM is smaller than any of the individual standard deviations, showing that the spread in the averages is smaller than the spread in any of the original data sets. (By now, this should come as no surprise!) In the space below, record your calculated SDM and explain why it makes sense that the averages in the table above should have *less* scatter than any of the original data sets.

d. Complete the following statement. "If I perform another 50-point experiment, there is a 68.3% chance that my new *average* will lie within ... " (Think carefully about this question, as it gets to the heart of what the SDM actually tells us!)

e. Explain in your own words how the standard deviation ($\sigma_{sd}$) differs from the standard deviation of the mean (SDM).

f. Using the data above, what would you report as your best estimate for the counts/interval and its uncertainty? (Write down your result in the form: value ± uncertainty.)

Note that having to perform many separate experiments is an awful lot of work! Fortunately, there is a much easier way of determining the SDM that does not involve performing a whole series of experiments. As shown in Appendix A, one can use *error propagation* techniques to derive a simple formula for the standard deviation of the mean (SDM), which is sometimes denoted as $\sigma_{\langle x \rangle}$. The result is

$$\sigma_{\langle x \rangle} = \text{SDM} = \frac{\sigma_{sd}}{\sqrt{N}}$$

where $\langle x \rangle$ is the average of a set of measurements, $\sigma_{sd}$ is the standard deviation of the set of measurements, and $N$ is the number of measurements in the data set. This formula for the SDM simplifies things considerably because we do not need to measure a full set of averages. Instead, we simply make *one* set of $N$ measurements from which we determine the average and standard deviation $\sigma_{sd}$. To get the SDM, we then divide the standard deviation by the square root of the number of measurements that were made.

### 2.6.2. Activity: Calculating the SDM the Easy Way

**a.** To get some practice calculating the SDM the easy way, let's go back to the data table in the previous activity. Assume that we performed a single experiment (e.g., Run #1), so that we have a single average and a single standard deviation. Using these values, calculate the SDM the easy way and write down your best estimate for the measured counts/interval along with your estimated uncertainty (in the form: value ± uncertainty). Compare this result to your result in part (f) of the previous activity.

**b.** You should have found that the SDM calculated the easy way is reasonably close to the SDM calculated the hard way. These values won't be exactly equal, but they should be similar in size and both should be smaller than any individual standard deviation reported in the previous activity. It might seem confusing that the two ways of obtaining the SDM (the hard way and the easy way) do not produce exactly the same result, but given the nature of what the SDM tells us this non-equality should not be too surprising. A similar non-equality is evident in the averages and standard deviations found in the previous activity. To help clarify this point, go back to the data in the previous activity and calculate the SDM (the easy way) for each row of the data table (this goes quickly if using a spreadsheet).

| Run # | Mean | SDM |
|-------|------|-----|
| 1     |      |     |
| 2     |      |     |
| 3     |      |     |
| 4     |      |     |
| 5     |      |     |
| 6     |      |     |
| 7     |      |     |
| 8     |      |     |
| 9     |      |     |
| 10    |      |     |

**c.** Examining the table above, you will see that we performed this experiment 10 different times and ended up with 10 different means and 10 different SDMs. But given what you have learned so far, do these results seem to be consistent with each other? Explain briefly what it means for results to be *consistent* with each other.

---

### Aside: Confidence Intervals

We close this section by introducing the concept of a *confidence interval* (a term you may have heard before) and relating it to what we have just learned. It's important to understand that scientists are typically exploring situations that have *not* been investigated previously. Therefore, a reported result may be the first time such a measurement has been made. In such a situation, it is critical to provide some measure of uncertainty so that the scientific community knows how much to "trust" the measured value. One such measure of uncertainty is the SDM. Another measure is the confidence interval, which provides a measure of "confidence" that the true value of a quantity lies within a particular range of values. For example, it is common for a researcher to provide a 90% confidence interval, which means there is a 90% chance that the true value lies within this range.

Recall that the SDM tells us there is a 68.3% chance that a *repeated measurement* of $\langle x \rangle$ (we assume the measurement is an average) lies within $\pm \sigma_{\langle x \rangle}$ of this measured value. On the other hand, a 68.3% confidence interval tells us that there is a 68.3% chance that the *true value* we are trying to measure lies within a particular range. Although it sounds like these two intervals are the same—and in fact, most physicists treat them as if they are—there is a subtle technical difference between the two.

Suppose we make a series of measurements and calculate the average and SDM. As we know, there is a 68.3% chance that a repeated measurement (of the average) will lie within the interval $\langle x \rangle \pm \sigma_{\langle x \rangle}$, a 95.4% chance that a repeated measurement will lie within the interval $\langle x \rangle \pm 2\sigma_{\langle x \rangle}$, and a 99.7% chance that a repeated measurement will lie within the interval $\langle x \rangle \pm 3\sigma_{\langle x \rangle}$. Note that the interpretation here regards a *repeated measurement* (of the average). For the confidence interval, it turns out that there is a 68.3% chance that the true value lies within the interval $\langle x \rangle \pm 1.00\sigma_{\langle x \rangle}$, a 95.4% chance the true value lies within the interval $\langle x \rangle \pm 2.00\sigma_{\langle x \rangle}$, and a 99.7% chance the true value lies within the interval $\langle x \rangle \pm 2.97\sigma_{\langle x \rangle}$. As you can see, there is very little difference between the measurement uncertainty and the confidence interval (which is why they are often used interchangeably).

For completeness, the three most common confidence intervals are 90%, 95%, and 99%. In terms of the SDM, the 90% confidence interval is $\langle x \rangle \pm 1.64\sigma_{\langle x \rangle}$, the 95% confidence interval is $\langle x \rangle \pm 1.96\sigma_{\langle x \rangle}$, and the 99% confidence interval is $\langle x \rangle \pm 2.58\sigma_{\langle x \rangle}$.

### Making Use of Uncertainty Estimates

Although we spent time discussing the standard deviation and the SDM, it turns out that we will not regularly need these concepts in this course. It's not that these ideas are unimportant; on the contrary, specifying an appropriate uncertainty is extremely important when reporting new scientific results. But in this course we will be focusing our attention on learning the fundamentals of physics as opposed to performing measurements as accurately as we can. You should, of course, be careful when making measurements, and we may occasionally ask for an estimate of the uncertainty in your measurements. But we prefer you to think about the physics and not worry too much about measurement uncertainty.

## 2.7  PROBLEM SOLVING

### 2.7.1. Activity: A Counting Experiment

Suppose your group measures the number of spontaneous radioactive decay counts occurring in a particular time interval using the radiation sensor. You want to get a really nice histogram, so you run this experiment for a long time to collect 90,000 data points and obtain the results shown in Fig. 2.7.

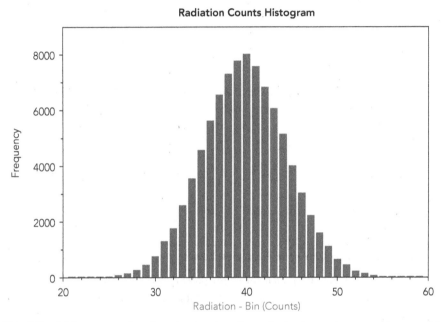

**Fig. 2.7.** A histogram showing the results of a radiation counting experiment with 90,000 data points.

**a.** *Estimate* the average (mean) and standard deviation for the data using the histogram. This is just an estimate based on the histogram.

**b.** If you were to take *one additional data point*, the result you get has a 68.3% chance of falling between what two values?

**c.** Calculate the (estimated) uncertainty in your *average value*.

**d.** If you were to collect data for twice as long (180,000 total points), how would you expect the *standard deviation* to change? In particular, would you expect the standard deviation to increase, decrease, or stay roughly the same? Explain briefly.

**e.** How would you expect the *standard deviation of the mean* to change when you take data for twice as long? In particular, would you expect the SDM to increase, decrease, or stay roughly the same? Briefly explain.

### 2.7.2. Activity: Different Radiation Sensors

Suppose your group takes a set of measurements from one of the radioactive sources using the sensor at your table. As before, you measure the number of decay counts in 1 second many times. You plot your data in a histogram, which produces the left set of bars ("Series 1") in Fig. 2.8. The group next to yours borrows the *same radioactive source* you were using and takes their own set of data using *the radiation sensor at their table* (placing the source the same distance from the sensor). Their data produces the histogram shown in the right set of bars ("Series 2").

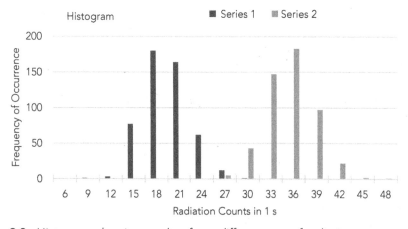

**Fig. 2.8.** Histogram showing results of two different sets of radiation measurements using two different radiation sensors.

    **a.** Do the two sets of data each show an *inherent uncertainty*? Briefly explain what this means and how you know.

    **b.** Would you say the two measurements *agree* regarding the number of decay counts in 1 second? In other words, are the measurements *consistent*? Carefully explain why or why not.

    **c.** What could explain why the two sets of measurements appear offset? Could this be due to an inherent uncertainty, or must it be something else?

# UNIT 3: INTRODUCTION TO ONE-DIMENSIONAL MOTION

*MrPreecha / Adobe Stock*

*Bats make use of echolocation to navigate with spectacular speed and agility. The process involves emitting a series of ultrasonic calls that echo back to warn them of obstacles. In this unit, you will use a similar technology to track the motion of objects to learn how position, velocity, and acceleration relate to one another.*

# UNIT 3: INTRODUCTION TO ONE-DIMENSIONAL MOTION

## OBJECTIVES

1. To describe one-dimensional motion using words, vectors, and graphs.

2. To acquire an understanding of position, speed, velocity, and acceleration and how these quantities are represented in graphs.

3. To understand the mathematical relationships between position, velocity, and acceleration and how to calculate these quantities from measurements.

## 3.1 OVERVIEW

A moving object might change its direction or speed as time passes. To accurately describe such movements, we must learn how to depict them both graphically and mathematically. The study of motion using mathematical equations and graphs is known as *kinematics*.

Describing an object's motion is not always easy. For example, a cloud in the sky could be changing its size and shape as it moves and attempting to describe such a complex motion can be overwhelming. To deal with such difficulties physicists often begin by making assumptions to simplify the situation. Therefore, we will begin our study of motion by concentrating on objects that are small and don't change their shape, treating the objects as mathematical points (even when they are not). In addition, we will focus on motion confined to a straight line, what we refer to as *one-dimensional motion*.

For the activities in this unit, we will use an ultrasonic motion sensor interfaced with a computer that measures the distance to an object. The program displays how an object's position changes over time in the form of a graph, and calculates other quantities such as velocity and acceleration. Our goal in this unit is *to develop an understanding of the concepts of position, velocity, and acceleration, as well as their formal mathematical definitions.*

## DESCRIBING MOTION WITH WORDS AND GRAPHS

### 3.2  POSITION CHANGES

The activities in this first section on kinematics will help us learn to describe changes in *position* using both *words* and *graphs*. Activities in the next section will involve descriptions of changes in the *velocity* of an object.

For the activities in this section, you will need:

- 1 ruler
- 1 ultrasonic motion sensor with attached computer interface

#### Motion Along a Line

In Section 1.8, we introduced the *position vector*, which is a vector whose tail is located at the origin and whose head points to some location in space. In the case of the solar system, the Sun is the natural location of the origin, and each planet can be specified by a three-dimensional (or two-dimensional) position vector relative to this origin. Of course, the planets move through space as time changes, so each position vector is actually a function of time as well. In fact, *position* and *time* are the two most fundamental measurements in the study of motion, and we will spend much of the next few weeks discussing them.

While a position vector is inherently three-dimensional, we can often work in a reduced number of dimensions. For example, in our solar system model, we assumed the planets all lied in a plane, in which case each planet could be specified by a *two-dimensional* position vector. In this next activity, we will consider a cart that is constrained to roll along a track. You will use a ruler to measure distances and specify the position of the cart shown in Fig. 3.1 relative to different coordinate systems. Each of the origins in the diagram is part of a Cartesian coordinate system having an *x*-axis that points horizontally to the right and a *y*-axis that points vertically up.

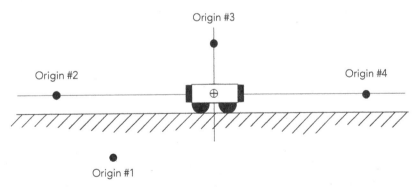

**Fig. 3.1.** A cart sitting on a track and various origins that can be used for finding position.

### 3.2.1. Activity: Finding Positions in Two Dimensions

**a.** Using a ruler marked in centimeters, find the position of the *center* of the cart relative to origin #1. That is, write down the two-dimensional position vector using vector notation (components and unit vectors $\hat{x}$ and $\hat{y}$) for this coordinate system.

**b.** Do the same thing for origin #2.

**c.** Do the same thing for origin #3.

**d.** Do the same thing for origin #4.

You should have found that with origin #1, the position vector had non-zero components in both the $x$ and $y$ directions. However, for origins #2, #3, and #4, one of the components was zero. A smart choice of coordinate system can simplify the problem, and for the situation of a cart on a track that can only move in the horizontal direction, the position of the cart is simplified by using a coordinate system in which the $y$-component is always zero (origin #2 or #4). If we choose one of these two origins, the situation is entirely *one dimensional*.

For one-dimensional motion, the position vector has only a single component and can be written $\vec{r} = r_x\hat{x}$ (or more commonly $\vec{r} = x\hat{x}$), where $r_x$ (or $x$) represents the position of the object along the $x$-axis (the $x$-component of the position vector).[1] It is important to remember that vector components can be positive or negative, depending on whether the object is positioned along the positive axis or the negative axis. Thus, the vectors $\vec{r} = -3$ m $\hat{x}$ and $\vec{r} = 2.5$ m $\hat{x}$ are both perfectly valid position vectors, the first indicating an object that lies 3 m from the origin along the negative $x$-axis, and the second indicating an object that is 2.5 meters along the positive $x$-axis.

Now for a further simplification. If we are dealing with one-dimensional motion and agree to always use the $x$-axis to describe the position of the object, we don't really need to continually specify that we're talking about the $x$-axis; we can simply say the object's position is $-3$ m or $+2.5$ m. In this case, there is no reason to write the unit vector $\hat{x}$, so instead we write only the *component* of the vector (the $x$-component, since we've agreed to always use the $x$-axis for one-dimensional motion). Thus, we might say that the object is at $x = -3$ m or $x = 2.5$ m. Note that when written in this form, we are not specifying the position *vector* but instead only the *x-component* of the position vector (a subtle but important distinction).

The standard convention for one-dimensional motion is to orient the positive $x$-axis so that it points to the right. With this convention, an object with a positive value of position indicates the object is located to the *right* of the origin, while a negative value indicates the object is to the *left* of the origin. However, this convention is not *always* followed, so you need to be mindful of the particular coordinate system being used when analyzing any problem.

---

### 3.2.2. Activity: Finding Positions in One Dimension

**a.** Consider the one-dimensional, horizontal system of the cart on the track shown in Fig. 3.1. Using your results from Activity 3.2.1, write down the *one-dimensional* position of the center of the cart relative to origin #2. Be sure to specify its sign (+ or −) and include units.

**b.** Similarly, write down the *one-dimensional* position of the center of the cart relative to origin #4. Again, be sure to specify its sign (+ or −) and include units.

---

[1] In general, the components of a vector $\vec{A}$ are written as $A_x$, $A_y$, and $A_z$, so consistency demands that the components of the position vector be written as $r_x$, $r_y$, and $r_z$. However, the variables $x$, $y$, and $z$ are (almost) always used to represent the components of the position vector: $r_x = x$, $r_y = y$, and $r_z = z$.

In Activity 3.2.1 we specified the position *vector*, while in Activity 3.2.2 we specified the position using only the *x-component* of the position vector. As long as the coordinate system has been specified in advance, an object's position can be given without having to use vector notation. Throughout the rest of this unit, we will consider one-dimensional motion using a coordinate system with a horizontal *x*-axis and the positive direction pointing to the right. So, unless specifically requested, you can drop the vector notation, saving it for when we consider motion in more than one dimension!

### The Ultrasonic Motion Sensor

In the rest of the activities in this unit, we will be using an ultrasonic motion sensor interfaced to a computer (see Fig. 3.2). The motion sensor acts somewhat like the bat shown at the beginning of the unit: it sends out a series of ultrasonic pulses (with a frequency that's too high to hear) that reflect from objects in the vicinity of the motion sensor. Some of the sound energy returns to the sensor, and by recording the time it takes for these reflected sound waves to return, the computer can use the speed of sound to calculate the distance to the object. There are a few things to be aware of when using a motion sensor.

### When Using a Motion Sensor

1. Do not get closer than about 0.2 m from the sensor because it cannot record reflected pulses that come back too soon after they are sent. If you get too close you may notice the recorded position suddenly jumps to an unexpected value. Similarly, the motion sensor might have trouble if you get farther than a few meters away.
2. The ultrasonic waves spread out in a cone of about 15° as they travel, and they will "see" the closest object. Therefore, be sure there is a clear path between the object you are tracking and the motion sensor (see Fig. 3.3). If your motion sensor loses track of the object, your graph will look nonsensical with big jumps in position; such situations are not uncommon, and you should be alert to such "garbage" data.
3. The motion sensor is very sensitive and can detect slight motions. You can try to walk smoothly along the floor, but don't be surprised to see small bumps in the graphs representing your steps (as we will see, such bumps are more noticeable in velocity and acceleration graphs).
4. Loose clothing like bulky sweaters are good sound absorbers and may not be "seen" very well by a motion sensor. If needed, you can hold a rigid, flat object to reflect the sound waves back to the motion sensor.
5. When using a motion sensor to measure an object's position, the motion sensor is usually assumed to sit at the origin and point along the positive *x*-axis. With this default setting, the motion sensor only measures positive positions. Although it's possible to override this default setting using the software, we will seldom have a reason to do so.

**Fig. 3.2.** One of several models of the type of motion sensor used in this Activity Guide. Image credit: Vernier.

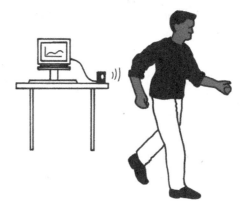

Fig. 3.3. Walking in front of a motion sensor attached to a computer.

**Important note on saving your files:** You may be prompted to save your data from a particular activity for use at a later time. When this occurs, you should save the graphs and associated data to a location you can access, perhaps labeling the files using the activity number.

### Position-Time Graphs of Your Motion

The purpose of the next activity is to learn how to relate graphs of one-dimensional position as a function of time to the motions they represent. How does the graph look when moving slowly or quickly? What happens when moving toward or away from the motion sensor? Typically, when plotting position versus time, the position of the object will be plotted as the dependent variable (on the vertical axis), while the time will be considered the independent variable (on the horizontal axis). For simplicity we will refer to such a graph as a *position-time* graph, but keep in mind that we are only plotting the one-dimensional position along a line (the *x-component* of the position vector).

To do the activity below and those that follow, you should make sure that: (1) the motion sensor is plugged into the interface, (2) the interface is on and connected to your computer, and (3) that the graphing software has correctly identified the sensor. If pre-configured experiment files are available and you wish to use them, look for the file named <A030203 (Position Graphs)> (or similar). Otherwise, you can typically set up the software to record a position-time graph for 10 seconds. Your instructor (or the Help file for your data acquisition software) may have specific instructions for how to accomplish this.

---

### 3.2.3. Activity: Interpreting Position Graphs

**a.** Try out different types of motion and observe what the position-time graphs look like. For example, try standing still, or walking *steadily* toward (or away) from the motion sensor at different speeds. Use the axes below to sketch three different graphs, providing a description of your motion (using words) for each case.

Sketch of graph                           Description of motion

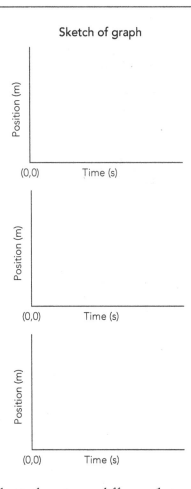

**b.** What is the primary difference between a graph made by walking *slowly* (and steadily) away from the motion sensor and one made by walking away *more quickly*?

**c.** What is the primary difference between a graph made by walking *toward* the motion sensor and one made by walking *away* from the motion sensor?

**Predicting the Shapes of Position-Time Graphs**

A good way to verify that you know how to interpret position-time graphs is to *predict* what the graphs will look like for a set of motions. Then you can carry out the experiments to test your predictions.

### 3.2.4. Activity: Predicting a Position-Time Graph

**a.** Suppose you stand 1.0 m in front of a motion sensor and walk steadily away for 2 seconds, stop for 4 seconds, and then walk toward the sensor more slowly so that it takes 4 seconds to end up back where you started. Sketch your prediction for the position-time graph on the following axes using a *dashed* line.

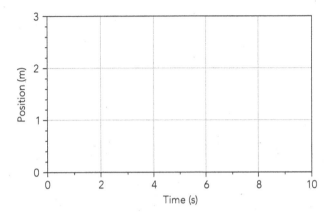

**b.** Now test your prediction by moving as described in part (a) while the motion sensor records your movement. Sketch the trace of the actual motion using a *solid* line on the above graph.

**c.** Is your prediction the same as the measurement? If not, describe how you would change either your movement or your prediction so they agree with the description of the motion. **Note**: It is unlikely that your prediction and the experiment will match *exactly*, even though the basic shapes may be roughly the same.

### Matching Position-Time Graphs

Let's now turn the activity you just did inside out. We'd like you to look at a position-time graph and then describe the motion using words. Then you will try to reproduce the motion yourself.

### 3.2.5. Activity: Matching Position-Time Graphs

**a.** Describe the motion depicted in Fig. 3.4 using words. Your description needs to specify precisely how someone should move to reproduce the above graph.

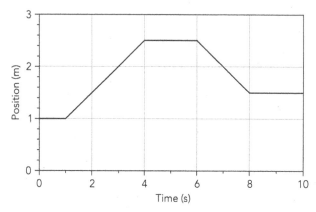

**Fig. 3.4.** Sample graph depicting a combination of standing still with two different types of motion.

**b.** Now try to reproduce the position-time graph above by moving in front of the motion sensor. You may need to try this a few times to get it right (though it's unlikely it will ever be a *perfect* match). Have different people in your group try to match it and sketch the best attempt on the graph below.

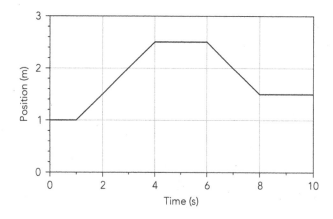

**c.** Which parts of the graph are the most difficult to accurately match? Why do you think this is the case?

---

### More Complex Position-Time Graphs

So far we have focused our attention on position-time graphs that consist of straight-line segments. However, you've probably noticed that the graphs you produced have small portions that are curved. Several examples of curved position-time graphs are shown in Fig. 3.5. The following activity explores what kind of motion is needed to produce such graphs.

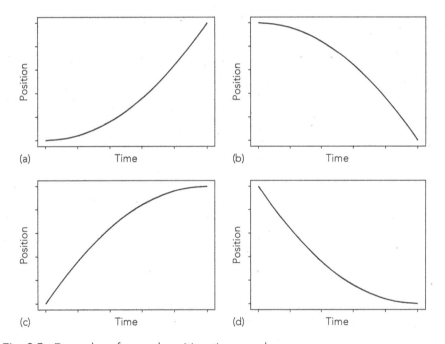

Fig. 3.5. Examples of curved position-time graphs.

---

### 3.2.6. Activity: Curved Position-Time Graphs

**a.** Consider the position-time graphs shown above. Describe how you would move to produce a graph with each of the shapes shown.

Graph (a) description:

Graph (b) description:

Graph (c) description:

Graph (d) description:

**b.** Now try to create curved position-time graphs like those shown by walking in front of the motion sensor. You don't need to match the graphs exactly; simply start with a blank graph and try to produce the basic shapes shown in Fig. 3.5.

**c.** What is the primary difference between motion that results in a *straight-line* position-time graph and motion that results in a *curved* position-time graph?

## 3.3  DESCRIBING VELOCITY WITH WORDS AND GRAPHS

Hopefully, you now have a reasonable understanding of position-time graphs, so we will turn our attention to something new: velocity.

For the activities in this section you will need:

- 1 ultrasonic motion sensor interfaced to a computer

### Describing Velocity Graphically

At this point you have created many position-time graphs for your motions, and you probably realize that there is a relationship between the speed of an object and its position-time graph. In fact, we will be interested in both the object's speed and its direction (something that can be described by a vector).

The term *velocity* is used to represent both the speed and direction of motion of an object.

Figure 3.6 shows a hypothetical position-time graph for someone driving a car on a trip. The following activity will help us determine the precise relationship between an object's position graph and the object's velocity. Once we know this, we will be able to plot the object's velocity as a function of time.

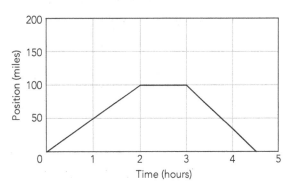

**Fig. 3.6.** An idealized graph depicting a car moving at a steady speed, standing still, and then moving at another (steady) speed in the opposite direction.

### 3.3.1. Activity: Creating a Velocity-Time Graph from a Position-Time Graph

**a.** During the first two hours of the trip, the car travels a distance of 100 miles in the positive direction. Using this information, determine the average speed of the car during this portion of the trip. Next, calculate the slope ("rise over run") of the position-time graph for the first two hours of the trip (be sure to include units in your calculation). How does the slope compare to the average speed of the car during this two-hour time period?

**b.** Determine the speed of the car during the next segment of the trip (between hour 2 and hour 3). Calculate the slope of the graph for this time period (again, be sure to include units in your calculation).

**c.** Lastly, determine the speed during the final segment of the trip. Calculate the slope of the graph for this time period. How does the slope compare to the speed you just calculated? What do you think the negative sign is telling us?

**d.** Notice that the slopes you calculated above provide both *speed* and *direction* information and therefore represent the *velocity* of the object. Use the slope information you calculated above to create a velocity-time graph for the car.

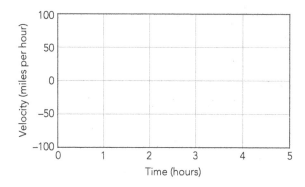

### The Definition of Velocity

Physicists use the concept of *velocity* to represent both the *speed* and *direction* of an object. Like position, velocity is inherently a vector quantity because it has both a magnitude (the speed) and a direction. For one-dimensional motion, the velocity is given by the slope of the corresponding position-time graph. For example, for motion along the $x$-axis we can write the $x$-component of the velocity mathematically as $v_x = \frac{\Delta x}{\Delta t}$, where the symbol $\Delta$ (the uppercase Greek letter delta) is interpreted to mean the *change* in a particular quantity (the final value minus the initial value). In this case we would write $\Delta x = x_2 - x_1$ and $\Delta t = t_2 - t_1$ so that $v_x = \frac{x_2-x_1}{t_2-t_1}$, which represents the slope of a straight-line segment between points 1 and 2.

As long as the position graph is a straight line, we can choose any two points on the line and get the same value for the velocity. However, if the position graph is curved (or a series of straight-line segments with different slopes), then the value of the velocity will depend on which two points we choose! While this may seem a little odd, the reason for this apparent inconsistency is that the formula $v_x = \frac{\Delta x}{\Delta t}$ actually represents the *average velocity* between the initial and final points. Unless the position graph consists of a single, straight line, the average velocity between two points will, in general, depend on the points you choose.

In addition to the average velocity, we may also want to know the *instantaneous* velocity, defined as the slope of the line *tangent* to the graph at a particular time. Mathematically, the instantaneous velocity is given by the *derivative* of the position-time graph (the slope between two points that are infinitely close together):

$$v_x = \lim_{\Delta t \to 0} \frac{\Delta x}{\Delta t} = \frac{dx}{dt} \tag{3.1}$$

### The Velocity Vector

When dealing with more than one dimension, an object's position should be considered a vector quantity, $\vec{r} = x\hat{x} + y\hat{y} + z\hat{z}$, so it becomes difficult to plot "the position" of an object in two (or three) dimensions. In these cases the concept of the "slope" of the position graph begins to lose its meaning. However, we can still define the average velocity to be the change in the position vector divided by the change in time:

$$\vec{v}_{\text{avg}} \equiv \langle \vec{v} \rangle = \frac{\Delta \vec{r}}{\Delta t} = \frac{\Delta x}{\Delta t}\hat{x} + \frac{\Delta y}{\Delta t}\hat{y} + \frac{\Delta z}{\Delta t}\hat{z} \tag{3.2}$$

where the bracket around $\vec{v}$ denotes the average. When written this way, we see that the vector nature of velocity is a direct result of the vector nature of position. As you might guess, the instantaneous velocity vector is defined as the *derivative* of the position vector with respect to time:

$$\vec{v} = \lim_{\Delta t \to 0} \frac{\Delta \vec{r}}{\Delta t} = \frac{d\vec{r}}{dt} = \frac{dx}{dt}\hat{x} + \frac{dy}{dt}\hat{y} + \frac{dz}{dt}\hat{z} = v_x\hat{x} + v_y\hat{y} + v_z\hat{z} \tag{3.3}$$

Stated another way, *velocity is the rate of change of position with respect to time*. The instantaneous velocity vector points in the direction the object is moving at that instant and, as with any vector, the magnitude of the velocity vector is given by its length, $v = |\vec{v}| = \sqrt{v_x^2 + v_y^2 + v_z^2}$.

As we saw in Unit 1, a vector can be represented by an arrow (whether in one or more dimensions). For velocity, the *length* of the vector represents the object's *speed*; the longer the arrow, the larger the speed. The *direction* of the arrow represents the *direction of motion*. Thus, if you are moving to the right, your velocity vector can be represented by an arrow that points to the right.

On the other hand, if you are moving twice as fast to the left, the arrow representing your velocity vector would be twice as long and point in the opposite direction.

Note that when using vectors, it does not make sense to talk about the vector being positive or negative. The direction of the vector is simply given by the direction the arrow is pointing, and this direction is independent of whatever coordinate system you are using. However, a vector's components *do* depend on the coordinate system being used, and for one-dimensional motion the vector's component provides information on the vector's direction (more on this in a moment).

### 3.3.2. Activity: Sketching Velocity Vectors

**a.** Assuming a motion sensor points to the right, sketch the velocity vector representing rapid motion *away* from the motion sensor. Would the motion sensor report a positive or negative value for this velocity (assuming the usual sign convention for the motion sensor)?

**b.** Sketch the velocity vector representing slower motion *toward* the sensor, half as fast as in part (a). Would the motion sensor report a positive or negative value for this velocity?

**c.** What does the velocity vector look like if you are standing still?

### Velocity in One Dimension

As we saw earlier for position, we do not need to use vector notation for one-dimensional motion and can instead simply make use of its one component. The same is true for velocity: for motion in one dimension along the $x$-axis both $v_y$ and $v_z$ will be zero, so the velocity vector reduces to $\vec{v} = v_x \hat{x}$. Because there is only one dimension, we can once again eliminate the unit vector and specify only the velocity component $v_x$ (which can be positive or negative, with units).

It is important to note that although we will refer to $v_x$ as "the velocity," this statement is technically incorrect because velocity is a vector and $v_x$ is merely one component of the velocity. However, for motion in one dimension the sign (+ or −) of $v_x$ indicates the *direction* of motion. If you move in the positive $x$-direction your velocity is positive; if you move in the negative $x$-direction your velocity is negative. The faster you move—in either direction—the larger the absolute value (the *magnitude*) of your velocity (the speed). As discussed, the ultrasonic motion sensor's axis is normally taken to be positive, and the sign of the velocity as reported by the sensor indicates the direction of motion relative to the sensor.

### Graphing Velocity versus Time in One Dimension

Graphs of velocity versus time are more challenging to create and interpret than those for position, so it can be helpful to create velocity-time graphs of your own motion. For the activities that follow you should open the data collection software and set it to plot velocity versus time. Scale the velocity axis from −1.0 to +1.0 m/s and the time axis to read 0 to 5 s (or use a preconfigured experiment file). If necessary, your instructor can help you display other types of graphs or change scales (or you can consult the Help file for your data collection software).

**Note**: Although the motion-recording software can display velocity-time graphs in real time, the motion sensor does not *directly* measure the velocity of the object. The motion sensor only measures the position of the object as a function of time; the velocity is then *calculated* numerically as the derivative of position with respect to time. Unfortunately, the process of numerically taking the derivative has a drawback: any small inconsistencies (bumps) in the position data will show up as larger inconsistencies (bumps) in the velocity graphs. The result is that velocity graphs are never as smooth as positions graphs.

### 3.3.3. Activity: Making Velocity-Time Graphs

a. Starting close to the motion sensor, try walking *away* from the motion sensor for a few seconds at a *steady speed* of about 0.5 m/s. If necessary, try this a few times until you get a velocity graph you're satisfied with and then sketch your result on the following graph. (Your sketch need not be a *perfect* replica of your graph, but you should draw the main features.)

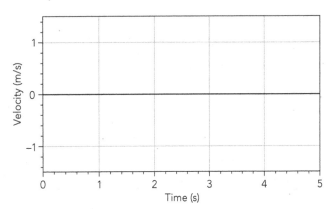

b. Now try walking *away* from the sensor for a few seconds at a *steady speed* of about 1.0 m/s. Sketch your graph below.

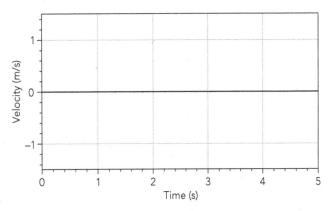

**c.** Next, walk *toward* the sensor at a *steady speed* of about 0.5 m/s. Sketch your graph below.

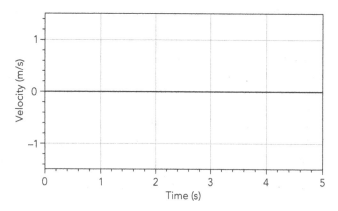

**d.** What is the key difference between the velocity graph made by walking away from the sensor *slowly* and the one made by walking away more *quickly?*

**e.** What is the key difference between the velocity graph made by walking *away* from the sensor and the one made by walking *toward* the sensor?

**Predicting Velocity-Time Graphs Based on Words**

Consider the following sequence of motions:

1. Stand still for 2 seconds.
2. Walk *away* from the sensor at a *steady speed* of about 0.5 m/s for 4 seconds.
3. Stand *still* for 3 seconds.
4. Walk *toward* the sensor at a *steady speed* of about 1.0 m/s for 3 seconds.
5. Stand still for 3 seconds.

### 3.3.4. Activity: Predicting a Velocity-Time Graph

**a.** Use a dashed line in the following graph to record your prediction for the shape of the velocity graph for the motion described above. Compare your prediction with your partner(s).

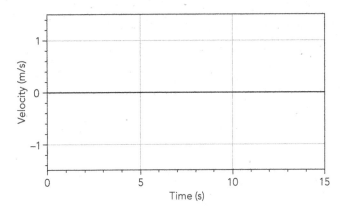

b. Now test your prediction. Practice moving in front of the motion sensor until you are confident that your motion matches the description in words above, and then draw the actual observed plot as a solid line on the axes above. Did your prediction match your actual motion? If not, what misunderstandings did you have?

c. Suppose the coordinate system used by the motion sensor is reversed, but the actual motion of the person walking remains the same. Use the axes below to sketch the graph in this case.

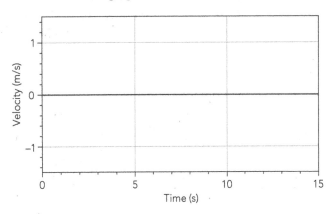

---

### Velocity Graph Matching

In the next activity you will try to move in such a way as to match a velocity graph shown on the screen (it can be quite challenging to match a velocity graph!). To do this activity, use the experiment file in your data collection software titled <A030305 (Velocity Match)>. The velocity graph in Fig. 3.7 should appear on the screen.

**3.3.5. Activity: Matching a Velocity Graph**

a.  Describe how you think you need to move in order to match the velocity graph in Fig. 3.7.

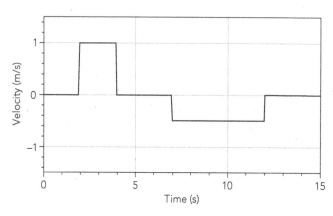

Fig. 3.7. Velocity graph to match.

b.  Now try to reproduce the graph experimentally. If you have trouble the first time, discuss it with your partners and try again. Make sure others in your group try it as well. Draw your group's best match on the axes above. Did your actual motion agree with your ideas in part (a)?

c.  You may have noticed it was difficult (if not impossible) to get the transitions between different velocities to match well. Is it possible for an object to move so that it produces a *perfectly vertical* line on a velocity-time graph? Explain briefly.

d.  The first time you tried this motion you may have run into the motion sensor on your return trip. If so, why did this happen, and how did you solve the problem? Whether or not you ran into the sensor, does a velocity graph tell you where to *start*? Explain briefly.

### 3.4  RELATING POSITION AND VELOCITY GRAPHS

**Creating a Velocity Graph from a Position Graph**

So far we have examined position-time and velocity-time graphs as different ways to represent an object's motion. But as we saw earlier, it is possible to figure out an object's velocity by examining its position-time graph. Conversely, it's also possible to determine how an object's position has *changed* by examining its velocity-time graph. In the following activities, we explore how position-time and velocity-time graphs are related.

You will need the following equipment for the activities in this section:

- 1 ultrasonic motion sensor interfaced to a computer

To complete the next activity, you should set the data collection software to display both position-time and velocity-time graphs *simultaneously* for a period of 5 seconds. Scale the vertical axes such that the position graph displays 0 to 4 m, while the velocity graphs displays −2 to +2 m/s.

#### 3.4.1.  Activity: Finding Velocity from Position

**a.** Carefully study the position-time graph shown below and *predict* the velocity-time graph that will result from the motion. Using a *dashed* line, sketch your prediction of the velocity-time graph on the velocity axes.

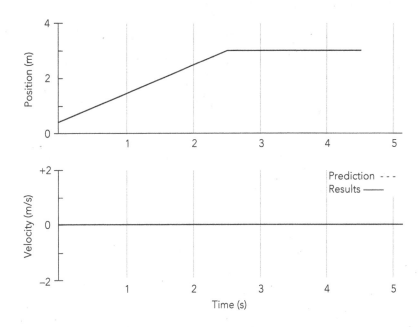

**b.** Test your prediction by using a motion sensor and walking to match the *position-time graph* shown. Make sure you obtain a good approximation of the position graph (you may need to try it a couple of times) and then sketch your actual position-time graph on the axes above. Once you are happy with the position graph match, use a *solid line* to draw the actual velocity-time graph. (Do not erase your prediction!)

How was your prediction? Briefly explain the reasons for any significant differences.

**c.** How would the *position* graph look different if you moved faster (or slower) during the initial segment?

**d.** How would the *velocity* graph look different if you moved faster (or slower) during the initial segment?

---

### Creating a Position Graph from a Velocity Graph

Our next goal is to understand how to produce a position-time graph from a velocity-time graph. As we will see, doing this requires us to know the position of the person or object for (at least) one specific time.

---

#### 3.4.2. Activity: Finding Position from Velocity

**a.** Carefully study the following velocity-time graph. Using a *dashed* line, sketch your *prediction* of the corresponding position graph on the top set of axes. (Note that we have *assumed* you are starting at a position of 1 m away.)

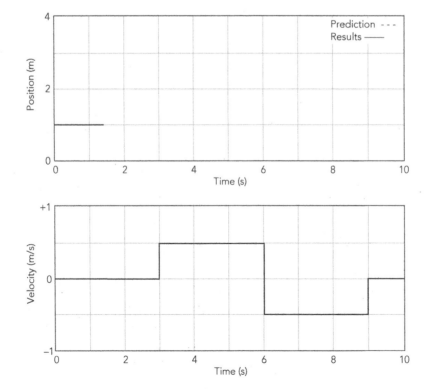

**b.** After sketching a prediction, try your best to duplicate the velocity-time graph by walking in front of the motion sensor, being sure to start at a position of 1 m. Try it a few times until you get a reasonable approximation of the velocity graph. When you are happy with your velocity-time graph, use a *solid line* on the top axes to draw the *actual* position-time graph. (Do not erase your prediction.) How was your prediction? Briefly explain the reasons for any significant differences.

**c.** How can you tell from a *velocity-time* graph that the moving object has *changed direction*? What is the velocity at the instant the object's direction changes?

**d.** We already noted that it is impossible to make perfectly vertical lines on a velocity-time graph. What about for a position-time graph? Is it possible to move your body (or an object) to make perfectly vertical lines on a position-time graph? Why or why not?

**e.** Recall that velocity is defined to be the derivative of position with respect to time. Does this relationship appear to hold for the graphs in this activity? Explain briefly.

## POSITION, VELOCITY, AND ACCELERATION

### 3.5  INTRODUCTION TO ACCELERATION

In addition to position and velocity there is a third quantity that is useful when describing the motion of an object—acceleration. Similar to how we defined velocity from position, acceleration is defined as *the rate of change of velocity with respect to time*. Mathematically, acceleration is the *derivative* of velocity with respect to time:

$$\vec{a} = \lim_{\Delta t \to 0} \frac{\Delta \vec{v}}{\Delta t} = \frac{d\vec{v}}{dt} = \frac{dv_x}{dt}\hat{x} + \frac{dv_y}{dt}\hat{y} + \frac{dv_z}{dt}\hat{z} = a_x\hat{x} + a_y\hat{y} + a_z\hat{z} \qquad (3.4)$$

Because of this definition, acceleration is a vector quantity, just like position and velocity.

In this section, you will examine the acceleration of objects. Unfortunately, because acceleration is *calculated* from velocity-time data, which is calculated from position-time data, acceleration graphs are not as smooth as velocity graphs. The result is that it is very difficult to use your body to produce a smooth enough acceleration graph for analyzing. For this reason, it is easier to examine the motion of a cart moving on a track.

For the activities in this section, you will need (Fig. 3.8):

- 1 ultrasonic motion sensor interfaced to a computer
- 1 low-friction cart
- 1 cart track, 2 m

**Fig. 3.8.** Schematic diagram showing a cart and motion sensor setup.

### Position and Velocity Graphs of a Low-Friction Cart After a Push

In the next two activities we will examine the velocity and acceleration of a low-friction cart *after* it has been pushed (after it has left your hand). Later, you will examine the acceleration of a low-friction cart *while* it is being pushed.

#### 3.5.1.  Activity: Cart Motion After a Push

**a.**  Based on your observations in earlier activities, *predict* how the position and velocity graphs will look if a low-friction cart is pushed *away* from the motion sensor and *released*. Assume the cart starts at the 0.5-m mark after it has been given a quick push so that it has an initial velocity of about 0.5 m/s. Sketch your predictions with dashed lines on the following axes, assuming that the data does not start recording until *after* the push is complete.

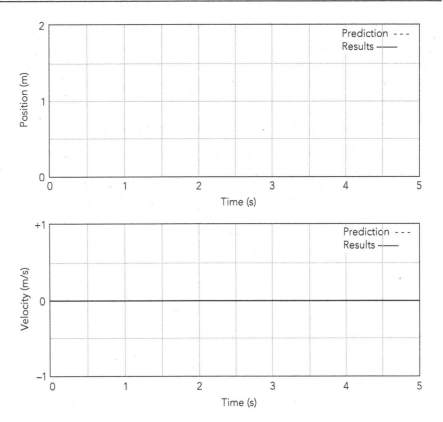

After you have made your predictions, set up the experiment as follows:

1. Place the motion sensor at the end of a (level) track. If the cart has a friction pad, move it out of contact with the track so that the cart can roll freely.

2. Open the motion software and display both position and velocity graphs.

3. When performing the experiments, give the cart a quick push away from the motion sensor and start collecting data *after* you let go of the cart. **Note:** It can be tricky to get the timing correct. If you're having trouble, it's okay if the sensor "sees" the push, but you should ignore this region when analyzing.

**b.** Collect data of the moving cart with the motion software. You may need to try it a few times to get a good result (you may only get 1–2 seconds of clean data). Sketch your results with *solid lines* on the axes above.

**c.** Did your position-time and velocity-time graphs agree with your predictions? If not, explain any inconsistencies. In particular, did you get the initial position correct?

**An Acceleration Graph Representing Constant Velocity**

In the previous activity, you should have observed that the *velocity* of a low-friction cart *after* it has been pushed is approximately *constant* (although you might notice that the cart slows down slightly due to friction). In the next activity, we will consider the acceleration for this motion.

**3.5.2. Activity: Graphing Acceleration versus Time**

a.  What should an acceleration-time graph look like for the cart motion you just observed? Use the definition of acceleration to sketch a dashed line on the axes that follow. **Note**: We are only interested in the smooth, constant velocity portion of the graph after the quick push.

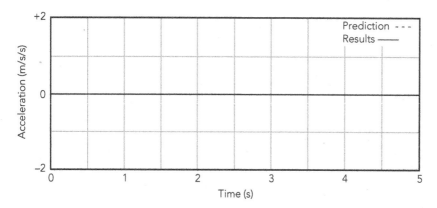

b.  In the motion sensor software, add an acceleration-time graph for the data in Activity 3.5.1. (Alternatively, you can change the graph axes to display the acceleration of the cart instead of the position.) Sketch the acceleration graph using a solid line on the axes above. Does the acceleration-time graph you observed agree with your prediction? In particular, what characterizes constant velocity motion on an acceleration-time graph? **Note**: You do *not* need to take any new data for this part!

Notice that in the previous activity the acceleration of the cart is (approximately) zero. Given the definition of acceleration shown in Eq. (3.4) and the fact that the cart's velocity is (approximately) constant, this should not be too surprising. In fact, the (one-dimensional) acceleration is given by the *slope* of a velocity-time graph.

**Finding Accelerations Using Motion Diagram Vectors**

As previously discussed, for one-dimensional motion it is not necessary to use vector notation for position or velocity, and the same holds true for acceleration. If we choose our coordinate system so the motion is entirely along the $x$-axis, then the position is described by $x$, the velocity by $v_x$, and the acceleration by $a_x$ (all of which can be positive, negative, or zero).

To find the *average* acceleration of the cart during some specific time interval, we need to measure its velocity at two different times, calculate the difference between the final value and the initial value, and divide by the time interval. For example, if a car changes from a velocity $v_{1x}$ at time $t_1$ to a velocity $v_{2x}$ at time $t_2$, then the change in velocity is given by $\Delta v_x = v_{2x} - v_{1x}$. The average acceleration over this time interval is then given by

$$\langle a_x \rangle = \frac{\Delta v_x}{\Delta t} = \frac{v_{2x} - v_{1x}}{t_2 - t_1}$$

where the bracket around $a_x$ denotes the average. Notice that this formula is nothing more than the slope of the line connecting two points on a velocity-time graph.

Although vector notation is not required in one dimension, it is sometimes helpful to think of acceleration using a vector picture. As shown in Fig. 3.9, a car is moving to the right and speeding up between times $t_1$ and $t_2$. The initial ($\vec{v}_1$) and final ($\vec{v}_2$) velocity vectors are shown above the car at times $t_1$ and $t_2$. To find the vector $\Delta \vec{v} = \vec{v}_2 - \vec{v}_1$ representing the *change in velocity*, we need to subtract (vectorially) the initial velocity $\vec{v}_1$ from the final velocity $\vec{v}_2$.

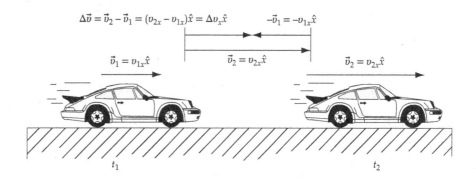

**Fig. 3.9.** Using a motion diagram and vector subtraction to find the change in velocity $\Delta \vec{v}$ of the car.

To subtract one vector from another, we simply add its negative (changing the signs of a vector's components will result in a vector with the same magnitude but pointing in the opposite direction). This process is outlined in the vector diagram at the top of Fig. 3.9, where we have redrawn vectors $\vec{v}_2$ and $-\vec{v}_1$ so that we can add them using the head-to-tail method. The result of this procedure is the vector $\Delta \vec{v}$. Lastly, to find the average acceleration over this time interval we divide $\Delta \vec{v}$ by the time interval $\Delta t$ (note that since $t_2 - t_1 > 0$ the acceleration vector will always point in the same direction as $\Delta \vec{v}$).

In this example, both the initial and final velocity vectors are pointing to the right, which means both $v_{1x}$ and $v_{2x}$ are *positive* quantities (assuming the positive $x$-axis points to the right). In addition, because $|\vec{v}_2| > |\vec{v}_1|$ (the car is speeding up), we know that $|v_{2x}| > |v_{1x}|$, which means $\Delta v_x = v_{2x} - v_{1x} > 0$. The fact that $\Delta v_x$ is *positive* tells us that the vector $\Delta \vec{v}$ points to the right (in the positive $x$-direction). (Of course, we already determined this in the vector diagram above, but it is helpful to understand both perspectives.)

### 3.5.3. Activity: Using Vectors to Find Acceleration

**a.** Imagine a scenario similar to the one above in which both initial and final velocity vectors are pointing to the right, but with $|\vec{v}_1| > |\vec{v}_2|$. In the space below, draw the vectors $\vec{v}_1$ and $\vec{v}_2$ and show how to calculate the vector $\Delta\vec{v}$. Does $\Delta\vec{v}$ (and hence $\vec{a}$) point to the right or to the left? In addition, explain whether the quantity $\Delta v_x$ (and hence $a_x$) is positive or negative. Finally, how would you describe the car's motion using words?

**b.** Now imagine another scenario: the initial velocity points to the right, but the final velocity points to the left, and $|\vec{v}_1| > |\vec{v}_2|$. Once again, draw the vectors $\vec{v}_1$ and $\vec{v}_2$ and show how to calculate $\Delta\vec{v}$. Does $\Delta\vec{v}$ (and hence $\vec{a}$) point to the right or to the left? In addition, explain whether the quantity $\Delta v_x$ (and hence $a_x$) is positive or negative. Would your answers change if $|\vec{v}_2| > |\vec{v}_1|$? Explain briefly.

## 3.6   VELOCITY AND ACCELERATION FOR CHANGING MOTION

In the previous activities we looked at position and velocity graphs for the motion of a cart moving at a (nearly) constant velocity. Our goal in this section is to examine the motion of a cart as its velocity *changes*.

For the activities in this section, you will need:

- 1 ultrasonic motion sensor interfaced to a computer
- 1 low-friction cart
- 1 cart track, 2 m
- 1 variable-speed fan for cart[2]

### Speeding Up Slowly

In the next activity we will examine graphs of position, velocity, and acceleration for a cart whose velocity is changing slowly. You should set up the cart, track, fan assembly, and motion sensor as shown in Fig. 3.10.

Keep the following points in mind when performing activities using the fan cart:

1. Always catch the cart with your hand at the end of a run before it crashes into the end stop.

---

[2] You can either use a variable-speed fan or a fixed-speed fan with a variable number of batteries (inserting conducting "dummy cells" in place of some batteries).

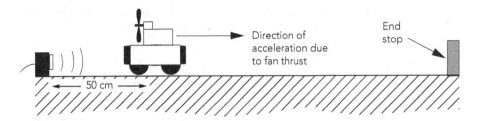

**Fig. 3.10.** Diagram of cart, fan assembly, motion sensor, and track with an end stop mounted on it. Note that the fan blade should not stick out beyond the end to the cart.

2. To keep the motion sensor from being affected by the fan rotation, be sure that the fan blades do not extend beyond the end of the cart. If it is an option, move the switch on the motion sensor to "cart mode."

3. Start by having the motion software display *Position* from 0.0 to 2.0 m and *Velocity* from −1.0 to +1.0 m/s for a total time of 3.0 s, as shown in the following graphs. If necessary, rescale the graphs so that the traces fill the screen.

4. You may need to try the experiment a couple of times to get clean data.

---

### 3.6.1. Activity: Speeding Up Slowly

**a.** *Predict* the shapes of the position-time and velocity-time graphs for a fan cart that starts at rest and moves away from the sensor. Sketch your predictions on the following axes using *dashed* lines.

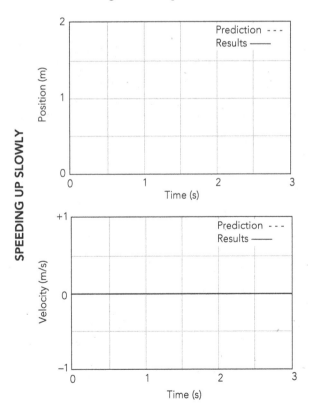

**b.** Turn on the fan to a *low speed* and release the cart from rest while collecting data to create position-time and velocity-time graphs of the fan cart as it moves away from the sensor. Sketch the graph representing the actual motion on the preceding axes using a *solid* line. How do the results compare to your predictions? Briefly explain any significant differences.

**c.** How does the *position* graph differ for this motion compared to steady (constant velocity) motion?

**d.** What feature of the *velocity* graph signifies that the motion is *away* from the sensor? What feature of the *velocity* graph signifies that the cart is *speeding up?* Does the velocity appear to be increasing at a steady rate or in some other way?

**e.** Add a new graph and plot acceleration versus time for the cart. (Alternatively, you can change the graph axes to display the acceleration of the cart instead of the position.) Adjust the scale as necessary so you can clearly see the portion of the graph where the cart is speeding up. Sketch your graph on the acceleration axes that follow.

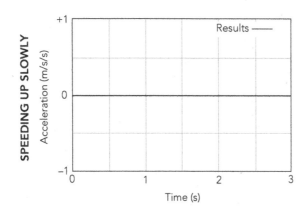

    **f.** During the time that the cart is speeding up, is the acceleration positive or negative? Does this indicate that the acceleration (vector) points to the right or to the left? Why does *speeding up* while moving *away* from the sensor result in this sign of acceleration (what is happening to the velocity)? If necessary, you may want to refer back to Fig. 3.9.

    **g.** Does the acceleration appear to vary as the cart speeds up, or does it seem reasonably constant? Is this what you would expect based on the velocity graph and the definition of acceleration? Explain briefly.

**Note**: You will want to compare these results with what you find in the next activity. This can typically be accomplished by "storing" the data from this experiment and then performing a new experiment in the next activity without exiting the file.

**Speeding Up More Quickly**

Suppose we increase the fan speed so the cart speeds up more quickly (see Fig. 3.11). How would your velocity and acceleration graphs change?

Fig. 3.11. Diagram of cart and fan assembly at high fan speed.

**3.6.2. Activity: Speeding Up More Quickly**

    **a.** Below, resketch the velocity and acceleration graphs you found in Activity 3.6.1 with the low-speed fan using a *solid* line. Then *predict* how the high-speed fan results might look using a *dashed* line.

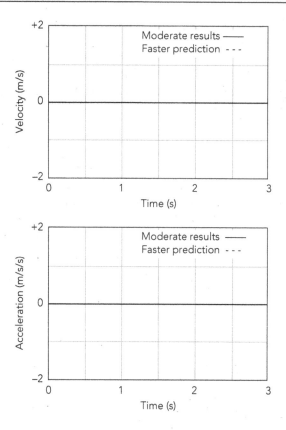

**b.** Test your predictions by accelerating the cart using a high-speed fan. Once you have clean data sketch the results on the graphs that follow.

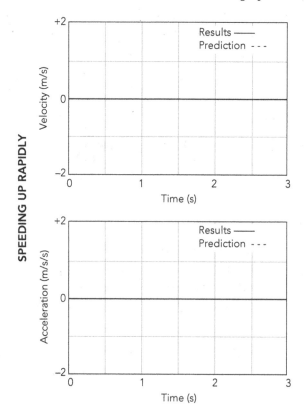

**c.** Do the general shapes of your velocity and acceleration graphs agree with your predictions? How has the acceleration graph changed compared to the previous experiment?

**d.** How has the velocity graph changed compared to the previous experiment?

## 3.7   CALCULATING VELOCITY AND ACCELERATION FROM POSITION DATA

By now you should have an intuitive feel for how to describe motion in terms of the position, velocity, and acceleration of an object. In this section we will use the definitions of average velocity and acceleration to calculate these quantities using position data. We will start by either recording position data of an accelerating cart or by analyzing a pre-recorded video.[3] The position data can then be used in subsequent calculations of velocity and acceleration.

### Using Video Analysis: Position as a Function of Time

To understand how the motion software translates (one-dimensional) position measurements into velocities and accelerations, it is helpful to start with the position and time data for a moving cart. Consider the motion of a cart with a fan like we studied in the last two activities. Suppose that instead of a motion sensor we use a video camera to film the cart. Each frame of the video allows us to determine the position of the cart at a different time by using video analysis software to step through the recording while marking the position of the cart in each frame. The frame rate of the video determines how much time passes between each frame.

To help visualize the process, Fig. 3.12 shows the results of a hypothetical video analysis. The cart began at position $x_1$ at time $t_1$ (corresponding to frame one of the video). The analysis consists of clicking on the location of the center of the cart in each frame of the video; these positions are marked by the dots in the figure. One can then extract the cart positions $x_1, \ldots, x_9$ at the corresponding times $t_1, \ldots, t_9$.

---

[3] The activity below describes how to obtain position data using video analysis. If you prefer to directly record the position data using a motion sensor, you can skip the video analysis instructions and start with the raw position data. But be sure to record at a relatively low data rate of around 2−3 points per second.

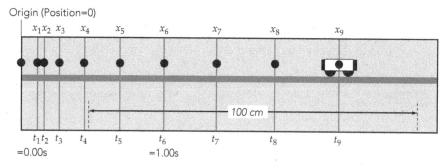

**Fig. 3.12.** A scale diagram of the position of an accelerating cart corresponding to nine different frames of a recorded video. A meter stick was placed in the plane of the cart motion for scale. At each time, the center of the cart is located at the position marked by the dot.

Figure 3.12 also shows three additional pieces of information necessary for the analysis: the origin (which defines the position $x = 0$), the time scale as set by the frame rate (frame 1 occurred at $t_1 = 0$ s and frame 6 occurred at $t_6 = 1$ s), and the distance scale (the distance between the points marked by dashed lines corresponds to 100 cm). From this information we can determine the position as a function of time.

### 3.7.1. Activity: Position-Time Data Using Video Analysis

a. Open the accelerating cart video <A030701 (Accelerating Cart Video).mp4> using your video analysis software.

b. Set the scale using the known length visible in the video; this tells the software how many meters are represented by each pixel in the video. (If necessary, you can enlarge the window containing the video for clarity.)

c. Collect position data for the moving cart, using the dot in the middle of the cart as your click-point in each frame. When finished, you should see a plot of the position of the cart as a function of time (the software automatically extracts the time between each frame based on the frame rate of the video). **Note**: To reduce the number of data points, you should set up the video analysis software to record position data twice per second.

### Determining Velocity and Acceleration over a Time Interval

We have seen that velocity is the rate of change of position, while acceleration is the rate of change of velocity. Thus, if we have determined the position of the cart at specific points in time, we should be able to calculate its velocity and acceleration using the position-time data. Specifically, if we measure the positions $x_1, x_2, x_3, \ldots$ at the respective times $t_1, t_2, t_3, \ldots$, then we can calculate the average velocity over each time interval. For example, the average velocity over the first time interval would be

$$\langle v_{1x} \rangle = \frac{x_2 - x_1}{t_2 - t_1}$$

Similarly, the average velocity over each subsequent time interval can be calculated as $\langle v_{2x} \rangle = \frac{x_3 - x_2}{t_3 - t_2}$, $\langle v_{3x} \rangle = \frac{x_4 - x_3}{t_4 - t_3}$, etc. Note that each of these velocities represents an *average* over the entire time interval. Thus, $\langle v_{1x} \rangle$ is the average velocity from $t_1$ to $t_2$ and is not simply the velocity at time $t_1$. In fact, it might be better to think of $\langle v_{1x} \rangle$ as the velocity halfway between $t_1$ and $t_2$ (more on this below). The average acceleration is calculated in a similar manner using the velocity values we just obtained. In particular, the average acceleration over the initial interval would be

$$\langle a_{1x} \rangle = \frac{\langle v_{2x} \rangle - \langle v_{1x} \rangle}{t_2 - t_1}$$

with the values over subsequent times following similarly.

Because we want to perform these calculations for many different positions, it would be beneficial to use a spreadsheet program. Figure 3.13 shows a sample spreadsheet that carries out these calculations. Column A lists the frames of the video, while columns B and C show the time and position data for each frame, as obtained from the video analysis program. (Note that a blank row is left between each frame to make Fig. 3.13 easier to understand, but you likely won't include these blank rows in the actual spreadsheet.)

| | A | B | C | D | E |
|---|---|---|---|---|---|
| 1 | Frame | Time (s) | x (m) | <v> (m/s) | <a> (m/s/s) |
| 2 | 1 | 0 | 0.139 | | |
| 3 | | | | = (C4-C2) / (B4-B2) | |
| 4 | 2 | 0.5 | 0.160 | | = (D5-D3) / (B4-B2) |
| 5 | | | | = (C6-C4) / (B6-B4) | |
| 6 | 3 | 1.0 | 0.209 | | |
| 7 | | | | | |
| 8 | 4 | 1.5 | 0.283 | | |
| 9 | | | | | |
| 10 | 5 | 2.0 | 0.384 | | |
| 11 | | | | | |
| 12 | 6 | 2.5 | 0.513 | | |
| 13 | | | | | |
| 14 | 7 | 3.0 | 0.668 | | |

**Fig. 3.13.** A sample spreadsheet program demonstrating how to calculate velocity and acceleration values from position and time data. Note that the blank rows are included to make the process easier to understand but should probably not appear in your spreadsheet.

Column D demonstrates how you could calculate the average velocity over each time interval following the equations above. For example, the mathematical expression in cell D3 performs exactly the calculation given in the expression for $\langle v_{1x} \rangle$ above: the change in position between frames 2 and 1 (C4–C2) is divided by the change in time between these same two frames (B4–B2). We located this value of the velocity *between* the two frames (in Row 3) since it is the average velocity over this time interval (between Frames 1 and 2). Once again, leaving a blank row in the spreadsheet is not necessary (or necessarily advised), it just helps visualize how the velocity is the average value between these two times. But keep in mind that if no blank rows are used, the formulas in columns D and E will need to be modified.

Of course, we need to repeat this calculation for all position and time intervals. Fortunately, a spreadsheet allows us to "click-and-drag" to quickly copy this formula down into the cells below, automatically updating the cell row numbers as needed. (The process of clicking and dragging to copy formulas into new cells is one of the most useful features of a spreadsheet program, so make sure you know how to do this.) In cell D5 we have shown how the formula will be updated when copied down to the cells below.

Notice that the velocity over any interval is calculated using the initial and final positions for that interval, so this procedure results in one fewer values of velocity than we have for position. Thus, although there are seven position values shown in the spreadsheet, there will only be six velocity values once we are finished (we essentially run out of position values). We will see some consequences of this fact in the next activity.

The acceleration calculation proceeds the same way, using adjacent pairs of velocity values. We will similarly end up with one fewer acceleration values than we have for velocity (and therefore two fewer values than position). In addition, you may notice that there is some ambiguity about what values of time to use in the average acceleration equation. Because the velocity values appear "in between" frames (it is the average velocity *over an interval*), these velocities do not occur at specific times. It might be most appropriate to assign times for the velocities that are halfway between those shown in column B. However, because the times are equally spaced, the time intervals separating the velocity values are the same as the time intervals separating the position values ($\Delta t = t_2 - t_1 = t_3 - t_2 = \cdots$).

In the next activity, we will calculate velocities and accelerations based on the position data taken from the accelerating cart video. To get started we will need to copy-and-paste the position and time data from the video analysis program into a spreadsheet program for analysis.

---

### 3.7.2. Activity: Calculating Velocity and Acceleration Manually

**a.** In the spreadsheet with your position data, create and fill in columns for velocity and acceleration, similar to those shown above. Remember that you can "click-and-drag" to copy a formula down after entering it the first time.

**b.** Create a position-time graph and describe its shape (straight, curved, flat, etc.). Does this make sense based on what is happening in the video?

**c.** Create a velocity-time graph and describe its shape (straight, curved, flat, etc.). Does this make sense based on what is happening in the video?

**d.** Lastly, create an acceleration-time graph and describe its shape (straight, curved, flat, etc.). Does this make sense based on what is happening in the video? (Your graph here might be quite "bumpy," but if you rescale the $y$-axis to include a larger range, say from 0 to $1\,\mathrm{m/s^2}$, it should look a little smoother.)

---

### Limitations of the Manual Calculation

The procedure you just used involves *two-point calculations* that come directly from our definitions of average velocity and average acceleration. In theory this should be perfectly fine, but the results may not look great when using experimental data. The issue is that the position measurements, like any measurements, are not perfect, and so the data "bounces" around slightly due mainly to inherent uncertainty (e.g., where, exactly, you clicked in the video). Although the position data may look quite smooth, it is not *perfectly* smooth.

Unfortunately, when you calculate the velocities these small errors in the position measurements become much larger errors in the velocity data. And these larger errors in the velocity data mean things get even worse when calculating the acceleration. You can observe this effect very clearly if you change one your position values by a small amount (say 10%) in your spreadsheet. This small change will hardly be noticeable on the position graph, but it will likely be noticeable on the velocity graph and even more noticeable on the acceleration graph.

A slightly more sophisticated technique typically leads to better (smoother) results. In practice, a software program will use more than two points (typically an odd number) to estimate the slope at a particular time. The more points that are used in the derivative calculation, the smoother the data will appear, but it comes at a cost. The downside of using more data points is that some real features of the data will get smeared out, particularly at the beginning and the end, so there is a tradeoff regarding how many points are best for a particular situation. (This is not something we will worry about in this course; we will simply use the default setting in our software program and then do our best to minimize errors when taking position data.)

## 3.8   INSTANTANEOUS VELOCITIES AND ACCELERATIONS IN ONE DIMENSION

As we have seen, instantaneous velocity is defined as the *derivative* of the position as a function of time, as given in Eq. (3.3). Similarly, instantaneous acceleration is defined in Eq. (3.4) as the derivative of the instantaneous velocity as a function of time. Thus, for an object moving in one dimension along the $x$-axis:

$$v_x(t) \equiv \frac{dx}{dt} \text{ and } a_x(t) \equiv \frac{dv_x}{dt}.$$

The "triple bar" symbol ($\equiv$) is used to show that this is how the instantaneous velocity and acceleration are *defined*.

Given these definitions, if we happen to know the position (or velocity) as a function of time, then the derivatives can be calculated quite easily.[4] In particular, consider the situation where the position as a function of time can be written as some power of time,

$$x(t) = bt^n + c$$

where $b$ and $c$ are constants, and $n$ a non-zero integer.[5] The instantaneous velocity is then given by the derivative as

$$v_x(t) \equiv \frac{dx}{dt} = nbt^{n-1}$$

Essentially, you bring down the exponent from the $t$, multiply it by whatever constant is in front, and then subtract one from the exponent. If there are multiple terms that involve powers of $t$, this same rule can be applied to each term.

For an example, suppose the position as a function of time is given by the function

$$x(t) = \left(5 \, \frac{m}{s^2}\right) t^2$$

where time is measured in seconds. Then the velocity as a function of time would be

$$v_x(t) = \frac{dx(t)}{dt} = \frac{d}{dt}\left[\left(5 \, \frac{m}{s^2}\right) t^2\right] = (2)\left(5 \, \frac{m}{s^2}\right) t^{2-1} = \left(10 \, \frac{m}{s^2}\right) t^1 = \left(10 \, \frac{m}{s^2}\right) t.$$

The acceleration is found by taking *another* time derivative, which also follows the power-law rule

$$a_x(t) = \frac{dv_x(t)}{dt} = \frac{d}{dt}\left[\left(10 \, \frac{m}{s^2}\right) t^1\right] = (1)\left(10 \, \frac{m}{s^2}\right) t^{1-1} = \left(10 \, \frac{m}{s^2}\right) t^0 = \left(10 \, \frac{m}{s^2}\right)$$

Note that although the units of the initial constant in the position function look somewhat funny for position ($5 m/s^2$), these units are necessary to make the equation dimensionally correct. In the position equation this constant is multiplied by $t^2$, which means it picks up units of $s^2$ in the numerator, cancelling the $s^2$ in the denominator and leaving the final units of position as meters (as they must be). Similarly, velocity ends up with units of m/s and acceleration with units of $m/s^2$.

### 3.8.1. Activity: Determining Velocity and Acceleration Using Differentiation

a.  Suppose $x(t) = (4m/s^2)t^2 + (3m/s)t$. First, find the general expression for $v_x(t)$. Then, determine the velocity at $t = 0$ s. Finally, determine the velocity at $t = 2$ s. (Don't forget units!)

---

[4] If you have seen the power-law derivative in a calculus class before, great. If you haven't, the power-law rule is quite simple and easy to use.

[5] Technically, $n$ does not need to be an integer, but all of our examples will assume it is.

**b.** Use your result from part (a) to find the general expression for $a_x(t)$. Then, determine the acceleration at both $t = 0$ s and $t = 2$ s (don't forget units!). Is the acceleration constant or does it change in time?

**c.** Now suppose $x(t) = (3\text{m/s}^4)t^4$. First, find the general expression for $v_x(t)$, and then determine the velocity at both $t = 0$ s and $t = 2$ s.

**d.** Use your result from part (c) to find the general expression for $a_x(t)$. Then, determine the acceleration at both $t = 0$ s and $t = 2$ s. Is the acceleration constant, or does it change with time?

∶• **3.9**  **PROBLEM SOLVING**

### 3.9.1. Activity: Cart Data

The position data for a fan cart moving along a horizontal track is shown below.

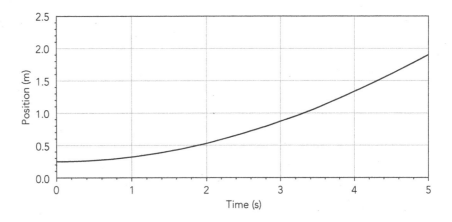

A good fit to the data is given by the function:

$$x(t) = \left(0.063\frac{m}{s^2}\right) t^2 + \left(0.022\frac{m}{s}\right) t + (0.24 \text{ m})$$

where $x$ is in meters and $t$ is in seconds. In the questions that follow be sure to show all of your work.

**a.**  Determine general expressions for $v_x(t)$ and $a_x(t)$. Is the velocity constant? Is the acceleration constant?

**b.**  Determine the position, velocity, and acceleration of the particle at the time $t = 2.0$ s.

**c.**  Use the definition of *average velocity* to find $\langle v_x \rangle$ of the particle over the interval $t = 0.0$ s to $t = 4.0$ s. Is this value the same as the instantaneous velocity at the middle of the interval (at $t = 2.0$ s)?

**d.** Similarly, use the definition of *average acceleration* to find $\langle a_x \rangle$ of the particle over the interval $t = 0.0$ s to $t = 4.0$ s. Is this value the same as the instantaneous acceleration at the middle of the interval (at $t = 2.0$ s)?

**e.** Based on the graph of the position data, the fit to this data, and your result from part (a), describe the motion of the cart in words. For example, how fast is the cart moving at the start? Is the cart speeding up or slowing down? And so on.

# UNIT 4: MOTION WITH CONSTANT ACCELERATION

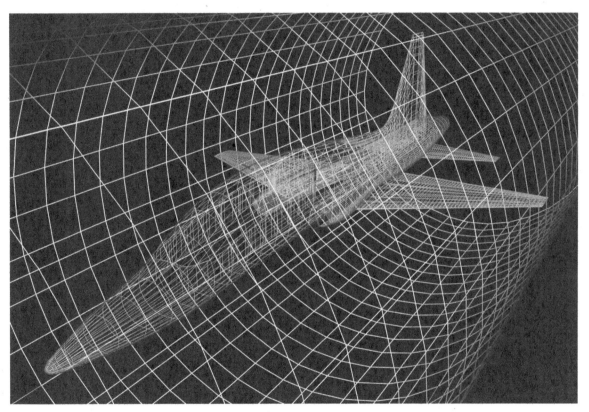

*chattereye / Adobe Stock*

*Physicists often describe objects or situations using models. A model can be a physical, scaled-down version of an object, a visual representation of such an object on a computer screen, or even a set of mathematical equations. For example, it is convenient to model the actual position and velocity of an object as a function of time using equations. Equations can be readily graphed, rearranged, and studied in ways that help illuminate the underlying motion of an object. In this unit, we will learn how to create mathematical models of objects moving with constant acceleration.*

# UNIT 4: MOTION WITH CONSTANT ACCELERATION

## OBJECTIVES

1. To recognize position-time, velocity-time, and acceleration-time graphs for constantly accelerated motion.

2. To use mathematical models to describe one-dimensional motion with constant acceleration.

3. To learn to use the *kinematic equations* for objects moving with constant acceleration.

## 4.1 OVERVIEW

In this unit, we will continue to study the motion of a low-friction cart propelled by a fan. In the first set of activities we will observe acceleration as the cart speeds up, slows down, and turns around. We then derive a useful set of formulas known as the *kinematic equations*, which describe how the position and velocity of an object change when undergoing a constant acceleration. These equations can be used to *model* the motion of our accelerating cart. A wide range of everyday motions involve constant (or near constant) acceleration, and we finish this unit by using the kinematic equations to describe the motion of various objects.

## SPEEDING UP, SLOWING DOWN, AND TURNING AROUND

### 4.2  DESCRIBING VELOCITIES AND ACCELERATIONS

#### Describing Velocities

The term *speed* is familiar to most students as a description of how fast or slow an object is moving. The speed can be calculated by taking the *distance* traveled and dividing by the time it takes to move this distance. The term *velocity*, on the other hand, is used less frequently and is calculated by taking the *change in position* and dividing by the time interval. Whereas the speed only tells us how fast the object moves, the velocity also tells us the direction of motion. For an object moving in one dimension, we saw that the change in position can be positive or negative, and this sign tells us the direction that the object is moving (in the positive *x*-direction or the negative *x*-direction).

Natursports/Shutterstock

   As we continue our study of motion we will find that the use of certain terms can lead to confusion. In particular, the terms *larger* and *smaller* can be ambiguous when describing velocities. For example, the term "larger velocity" could mean that an object is moving at a faster speed, or perhaps that the velocity is more positive (even though the object is moving slower). Thus, when comparing velocities, the terms greater than, less than, larger than, and smaller than should be used with caution (or avoided altogether). A less ambiguous way to compare velocities is to refer explicitly to the *speed* (magnitude of velocity) and *direction* (relative to the chosen coordinate system).

#### Describing Accelerations

When describing accelerations the same kinds of ambiguities can arise. However, there is not an analogous word like "speed" for the magnitude of acceleration. Therefore, when describing accelerations we will simply refer to the *magnitude of acceleration* as being larger or smaller, as well as specifying its *direction* (positive or negative for one-dimensional motion). In addition, you may have heard people use the term "deceleration" when an object is slowing down. The term deceleration can be ambiguous; in addition to slowing down, deceleration might mean the object has a negative acceleration (which could describe an object that slows down *or* speeds up). Because of this ambiguity, we will avoid using the term deceleration.

   In the activity that follows we use the definition of acceleration to determine the direction and magnitude of the acceleration for different one-dimensional situations (assumed to be along the *x*-axis).

---

#### 4.2.1.  Activity: Determining Average Acceleration

**a.** A runner's velocity changes from $-10\,\text{m/s}$ to $-6\,\text{m/s}$ in 1 second. Describe the runner's motion in words and determine both the magnitude and direction of the average acceleration during this time interval.

**b.** At the beginning of a 2-minute time period, a snail is moving at −1.0 mm/s. The snail then slows down, turns around, and starts heading back in the opposite direction at +0.2 mm/s. Determine the magnitude and direction of the average acceleration during this time interval.

**c.** A car's velocity changes from +20 mi/hr to +16 mi/hr in 3 seconds. Describe the car's motion in words and determine the magnitude and direction of the average acceleration during this time interval.[1]

**d.** A car's velocity changes from −20 mi/hr to −30 mi/hr in 2 seconds. Describe the car's motion in words and determine the magnitude and direction of the average acceleration during this time interval.

**e.** A donkey pulls a cart at +1 m/s and is still moving at +1 m/s an hour later. Describe the donkey's motion in words and determine the magnitude and direction of the average acceleration during this time interval.

## 4.3 ACCELERATION: SLOWING DOWN AND TURNING AROUND

To get a better feel for acceleration, it's helpful to examine velocity-time and acceleration-time graphs for different types of motion. As in Unit 3, we will use the motion sensor to observe a cart as it moves. We will begin by examining what happens when the cart is pushed away from the sensor, slows down, turns around, and then speeds up again.

---

[1] Because the velocity is given in miles-per-hour and the time interval in seconds, there is some question regarding the best units for the acceleration here. While miles-per-hour-squared or miles-per-second-squared may appear the most consistent, miles-per-hour-per-second may be the most useful. In any case, make sure you are careful with the units!

To complete the activities in this section, you will need the same apparatus you used in Unit 3; this includes:

- 1 ultrasonic motion sensor interfaced to a computer
- 1 low-friction cart
- 1 cart track, 2 m
- 1 fan assembly for cart

You should start the fan with its thrust pushing it toward the motion detector, and then give the cart a quick push *away* from the motion sensor. The cart should move away from the sensor, slow down, reverse direction, and then move back toward the sensor again (see Fig. 4.1). Try it a couple of times **before activating the motion sensor** to get a feel for how hard to push the cart. *Be sure not to let the cart crash into the motion sensor on its way back!*

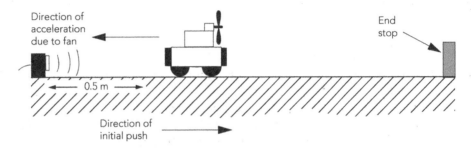

**Fig. 4.1.** A fan cart is set up such that after an initial push to the right (away from the sensor), the cart will slow down, turn around, and then speed back up toward the sensor.

---

### 4.3.1. Activity: Reversing Direction

**a.** For each part of the motion after the cart has left your hand—moving away from the sensor, at the point where it turns around, and moving toward the sensor—*predict* in the table below whether the velocity and acceleration will be positive, negative, or zero.

|              | Moving away | Turning around | Moving toward |
|--------------|-------------|----------------|---------------|
| Velocity     |             |                |               |
| Acceleration |             |                |               |

**b.** On the graphs below, sketch your predictions for the velocity and acceleration as a function of time for the *entire motion* (including the push). Your graphs don't need to be numerically accurate; we just want the basic *shapes* of the graphs.

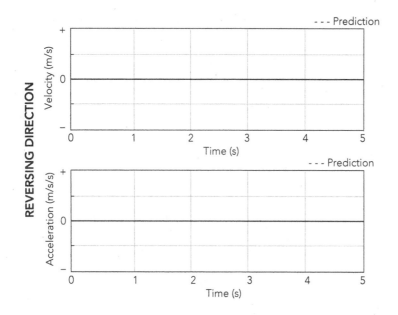

c. To test your predictions, open the motion software and display both velocity and acceleration graphs. Start the data and give the cart a quick push away from sensor. *You may need to try a few times to get a good run.* If necessary, adjust the scales on the graphs so you can see the plots clearly. When you get a good run, sketch both graphs on the axes below.

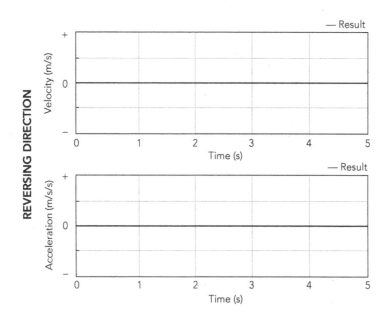

d. Label *both* velocity and acceleration graphs with the following points:
   • "A" when you start pushing the cart.
   • "B" when your push ends (when your hand stops touching the cart).
   • "C" when the cart has stopped and is about to reverse direction.
   • "D" when you catch the cart after its return trip.

e. Does the cart ever have zero velocity (check your velocity graph)? Does this agree with your prediction? How much time does the cart spend with zero velocity before it starts back toward the sensor? Explain briefly.

f. Describe the acceleration graph for the portion of the motion *after* the cart has left your hand and *before* you catch it (i.e., when it is moving on its own). Explain how the acceleration graph is related to the velocity graph.

g. What is the acceleration at the *instant* the cart is turning around? Is it positive, negative, or zero? Does this agree with your prediction? Explain why the acceleration has this value near the point where the cart turns around. **Hint:** Remember that acceleration is the *rate of change* of velocity.

h. Now set your motion software to display a *position-time* graph for this motion (you don't need to take new data, just add a new plot!). Use the space below to sketch your graph and add the points "A," "B," "C," and "D" to this plot, just like you did in part (d) for the velocity and acceleration graphs.

i. Describe the basic *shape* of the position-time graph when you are no longer touching the cart (after your push has ended but before you catch it). The graph might look like a function you've seen before. If so, what type of function does it look like?

---

**Note**: You will want to save this data for possible use in an upcoming mathematical modeling activity in Section 4.5.

Many people are surprised by the results of the previous activity. It is a common belief that the acceleration is zero at the point where the cart turns around. However, the experimental data should make it clear that the acceleration is constant once the cart has left your hand!

Now that you have some experience with the motion of the fan cart, let's look at a similar situation. In the following activity, you should start with the fan pointing so that its thrust is *away* from the motion sensor (opposite to the last experiment). You will then give the cart a quick push *toward* the motion sensor (make sure you start far enough away so that the cart doesn't get too close to the motion sensor). The cart will then move toward the motion sensor, slow down, reverse direction, and move away from the motion sensor.

### 4.3.2. Activity: Reversing Direction Again

**a.** For each part of the motion after the cart has left your hand—moving toward the sensor, at the point where it turns around, and moving away from the sensor—*predict* in the table below whether the velocity and acceleration will be positive, negative, or zero.

|              | Moving toward | Turning around | Moving away |
|--------------|---------------|----------------|-------------|
| Velocity     |               |                |             |
| Acceleration |               |                |             |

**b.** On the graphs below, sketch your predictions for the velocity and acceleration graphs for the *entire motion* (including your push). As before, your graphs don't need to be numerically accurate; we just want the basic *shapes* of the graphs. If you think carefully about the last experiment, you should be able to deduce what these graphs will look like.

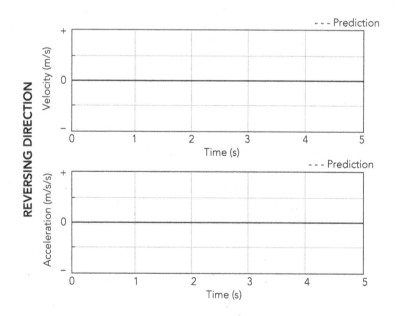

**c.** To test your predictions, open the motion software and display both velocity and acceleration graphs. Start the data collection and give the cart a quick push toward the sensor. *You may need to try a few times to get a good run.* If necessary, adjust the scales on the graphs so you can see the plots clearly. When you get a good run, sketch both graphs on the axes.

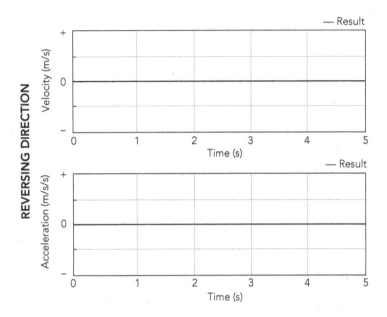

**d.** How were your predictions this time around? Note once again how the acceleration is related to the velocity, and in particular the values of velocity and acceleration at the instant the cart is turning around. Hopefully, the previous experiment helped you develop an understanding for this situation. For completeness, make a rough sketch of the position-time graph for this motion.

**Tossing a Ball**

Before moving on, it's worth considering one more situation that is quite similar to the experiments just performed. Suppose you throw a ball straight up into the air. In this situation the ball slows down as it moves upward, reaches a maximum height, and then speeds up again as it moves back down toward your hand. Notice how similar this motion is to the motion of the cart we just observed. Let's look at the ball's velocity and acceleration at various points during its motion.

### 4.3.3. Activity: The Rise and Fall of a Ball

**a.** Consider the ball toss experiment described above. Assume that *upward* is the positive direction. Using this coordinate system, does the ball toss look more like the experiment in Activity 4.3.1 or 4.3.2? Explain how the motion of the ball in this activity is like the motion of the cart.

**b.** Indicate in the following table whether you think the velocity and acceleration are positive, negative, or zero during each of the three parts of the motion (*after* the ball has been released). **Hint**: Use your knowledge of the fan cart experiment to guide you.

|  | Moving up | At highest point | Moving down |
|---|---|---|---|
| Velocity |  |  |  |
| Acceleration |  |  |  |

**c.** Make a rough sketch of what you think the position, velocity, and acceleration graphs will look like for the ball toss experiment. (We are only interested in the motion *after* the ball has left your hand.)

**d.** If time permits, try obtaining data for this situation by setting up a motion sensor on the floor (pointing upward) and then tossing a basketball above the motion sensor (being sure to catch the ball before it hits the motion sensor). Alternatively, you can use the pre-recorded experiment file <A040303 (Bowling Ball Toss)>. In either case, look at the position, velocity, and acceleration graph and comment on what you observe. If the acceleration appears to be constant, determine its value by fitting a straight line to the velocity-time graph and report this value below.[2]

---

[2] Although you might be tempted to use the acceleration graph to determine the acceleration, remember that the acceleration graph is subject to more uncertainty due to how it is calculated from the velocity data. Therefore, it's usually best to obtain the acceleration by fitting a line to the middle portion of the velocity graph (the end portions of the velocity graph are not as accurate because of the way velocity is calculated from the position data and due to air resistance).

The similarities between the motions of the fan cart and the ball suggest that the acceleration of the tossed ball will be constant. However, we must realize that the physical mechanisms responsible for these two motions are quite different. In one case a fan is responsible for causing the cart to accelerate, while in the other case it's a gravitational force. Because these mechanisms are different, we can do little more at this point than speculate about the motion of the ball and must turn to experiments for the answer (physics is an *experimental* science at heart).

You should have observed that the acceleration of a tossed bowling ball *is* (essentially) constant, just like the acceleration of the fan cart. We will explore the motion of objects near the surface of Earth much more thoroughly in a later unit. For now we simply note that we have seen two situations in which the acceleration is observed to be constant. Although most everyday motions do not exhibit constant acceleration, it is a common enough occurrence to devote some time to examining this specific type of motion.

## 4.4 EQUATIONS DESCRIBING CONSTANT ACCELERATION

As we have seen, quantitative information about velocity and acceleration is contained in a position-time graph. Similarly, the equations describing position, velocity, and acceleration are also related. In this section we derive the equations that describe the motion of an object moving with constant acceleration (including the possibility of zero acceleration).

### Velocity for Constant Acceleration

We previously discussed the concept of average acceleration, or the change in velocity over a specified time interval. For one-dimensional motion, we need only concern ourselves with the $x$-components of the vectors (assuming the motion is in the $x$-direction), so the average and instantaneous accelerations are given by:

$$\langle a_x \rangle = \frac{\Delta v_x}{\Delta t} = \frac{v_{2x} - v_{1x}}{t_2 - t_1} \text{ and } a_x = \lim_{t \to 0} \frac{\Delta v_x}{\Delta t} = \frac{dv_x}{dt} \tag{4.1}$$

where the subscripts 1 and 2 indicate arbitrary initial and final points, respectively. Normally, the instantaneous acceleration is different from the average acceleration; however, a *constant* acceleration always has the same value, so there is no distinction between the two: $\langle a_x \rangle = a_x$. Thus, for constant acceleration we can refer to $a_x$ simply as *the* acceleration.

### 4.4.1. Activity: Deriving the Velocity Equation

**a.** Let's assume we start monitoring the motion of an object at some initial time $t_0$. (In physics it is common to use $t_0$, pronounced "*t*-zero" or "*t*-naught," as the initial time.) The initial velocity in the $x$-direction at this time is then labeled as $v_{0x}$ ("*v*-zero" or "*v*-naught"). Similarly, we will label the final velocity at time $t_f$ as $v_{fx}$. Starting with the equation $a_x = \Delta v_x / \Delta t$, show (mathematically) that the final velocity can be written as $v_{fx} = v_{0x} + a_x(t_f - t_0)$ (this is straightforward—don't overanalyze!).

**b.** It is often the case that the initial time occurs when we start our stopwatch, so that $t_0 = 0$. In addition, the final time $t_f$ can take on *any* value; it's completely arbitrary. Thus, instead of writing it as $t_f$, which looks like a specific time, we will use the variable $t$ to represent *any* time in the future. The velocity at this arbitrary time would then be written *as a function of time*: $v_{xf} = v_x(t)$. Make these changes to write down a (slightly) simplified version of the above equation (again, don't overanalyze).

$$v_x(t) =$$

**c.** We now have an expression for the instantaneous velocity as a function of time. If this equation is correct its derivative should be equal to the acceleration. Check that this is true by taking the derivative $dv_x/dt$ and showing that you get $a_x$. (If you need a reminder about power-law derivatives, refer to Section 3.8.)

---

Note that by simply starting with the definition of acceleration, we have determined an equation that gives us the (instantaneous) velocity at any time in the future, as long as the acceleration is constant. All you need to know is the initial velocity and the (constant) acceleration. This equation is extremely useful when solving problems with constant acceleration and is one of the three primary *kinematic equations*.

**Position for Constant Acceleration**

In the previous activity we started with the definition of average acceleration $\langle a_x \rangle = \Delta v_x / \Delta t$ to come up with a useful expression for the velocity as a function of time:

$$v_x(t) = v_{0x} + a_x t \tag{4.2}$$

Recall that there is a similar definition for the average velocity $\langle v_x \rangle = \Delta x / \Delta t$, so you might be wondering if we can follow the same procedure as above to derive an equation for the position as a function of time. Doing so leads to the equation

$$x(t) = x_0 + \langle v_x \rangle t,$$

where $x_0$ is the initial position at $t_0 = 0$. Unfortunately, unlike our previous analysis, we are *not* able to make the substitution $\langle v_x \rangle \rightarrow v_x$ because the velocity is *not* constant! However, because the velocity equation is linear in time (when the acceleration is constant), we can find the average velocity without too much difficulty.[3]

### 4.4.2. Activity: Deriving the Position Equation

**a.** The average of a linear function over a particular range is given by the average of the two endpoints. Thus, for a time range that runs from 0 to $t$, the average velocity can be written as $\langle v_x \rangle = \frac{1}{2}[v_{0x} + v_x(t)]$. Substitute this result into the equation $x(t) = x_0 + \langle v_x \rangle t$ and make use of Eq. (4.2) to show that the position is given by the quadratic function $x(t) = x_0 + v_{0x}t + \frac{1}{2}a_x t^2$.

**b.** Explain in words what the constants $x_0$, $v_{0x}$, and $a_x$ represent.

**c.** Now differentiate the position equation with respect to time to make sure you get back the velocity equation given in Eq. (4.2).

### Summary: The Kinematic Equations

We have found two of the primary kinematic equations valid when the acceleration is constant. If we also consider an (very simple) expression for the constant acceleration, we can write down equations for each of the three kinematic quantities:

$$a_x(t) = a_x \text{ (constant)} \tag{4.3a}$$

$$v_x(t) = v_{0x} + a_x t \tag{4.3b}$$

$$x(t) = x_0 + v_{0x}t + \frac{1}{2}a_x t^2 \tag{4.3c}$$

If we know the acceleration, the initial velocity $v_{0x}$, and the initial position $x_0$ (the "initial conditions"), we can find the velocity and position at any subsequent time.

**Note**: It is important to realize that the equations above are only valid for *constant* acceleration in the $x$-direction and when the initial time is equal to zero:

---

[3] If you're familiar with integral calculus you might recognize that since the velocity is defined as the *derivative* of position, we can find the position by *integrating* the velocity: $x(t) = \int v_x(t)\, dt$.

$t_0 = 0$. Although this is usually the case it is worth noting that when the initial time is not zero, the time $t$ in Eqs. (4.3) should be replaced by $\Delta t = t - t_0$, where $t_0$ is the (non-zero) initial time. The result of making this change is that all graphs will be shifted to the right by $t_0$ units.

## MATHEMATICAL MODELING OF MOTION

### 4.5   MODELING MOTION WITH A CONSTANT ACCELERATION

In the last section we found that for constant acceleration, the position is a *quadratic* function of time. In this section we want to create a quadratic model for the position data so that we can understand how the parameters affect the shape of the graph. For the following activity we need some position data for motion under a constant acceleration (results from a sample experiment are shown in Fig. 4.2). We made graphs of such motion earlier with the fan cart, but we want to analyze only the portion of the motion when the cart is moving on its own (without seeing the initial push). You can either select only the region after the push, or you can record new data where you give the cart a push before starting the software. (Your instructor may have pre-recorded data for you to use for this activity.)

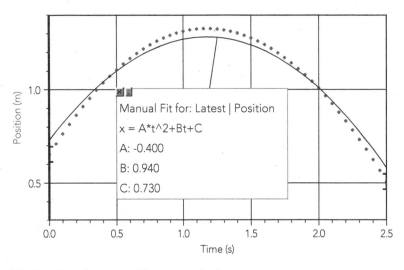

**Fig. 4.2.**  Results of a manual fit to sample data.

#### 4.5.1.  Activity: Modeling Position Data

**a.**   Once you have some position-time data on the screen, begin by *estimating* the initial position, the initial velocity, and the (constant) acceleration. Your estimates should arise simply by looking at the position-time graph; you should not make use of the computer here. **Hint**: Because the acceleration is constant, an estimate of the average acceleration over the entire interval is a reasonable approach.

**b.** Now we want to *model* the data, which we do by performing a *manual curve fit* with a quadratic function (a parabola) of the form $At^2 + Bt + C$. **Note**: You are *not* having the computer find the best fit for you! To start, use your estimates from part (a) to set the parameters $A$, $B$, and $C$. Then try adjusting the parameters manually to get the best fit you can. Print out (or sketch) a copy of your graph below.

**c.** Describe how the graph changes when you vary parameter $C$. Does this make sense based on what this parameter represents physically?

**d.** Describe how the graph changes when you vary parameter $B$. Does this make sense based on what this parameter represents physically?

**e.** Describe how the graph changes when you vary parameter $A$. Does this make sense based on what this parameter represents physically?

**f.** Although we can always adjust the parameters by hand, we can also ask the computer to find the parameters $A$, $B$, and $C$ that produce the best fit to the data. Have the computer perform its own *best fit* and write the equation below. How do the computer's fit parameters compare to yours?

**g.** Use the computer's best-fit equation for $x(t)$ to calculate the position at some time in the middle of the motion (e.g., 1.5 seconds), explicitly showing your calculation. How does this result compare to the value recorded directly in the data table? Is it exactly the same? Is it close?

---

The quality of the fit in the previous activity is an indication of how well Eq. (4.3c) describes the actual experiment. In theory, with no friction or air resistance, a quadratic model should fit extremely well. In reality the fit won't be perfect, particularly if the acceleration is less than about $1 \, \text{m/s}^2$. But even if the fit is not perfect, it should still provide a reasonable description of the motion.

We could repeat this process for the velocity and acceleration data, but it is easier to use the computer's fit parameters from the position data directly. This is the topic of the following activity.

---

### 4.5.2. Activity: Modeling Velocity and Acceleration Data

**a.** Change the plot so you are viewing a velocity-time graph. Then perform a manual fit of the velocity graph using the appropriate fit parameters from the computer's best fit to the *position* data. **Note**: You should now be using Eq. (4.3b) for your model! Print out or sketch a copy of your graph with model below.

**b.** Using your *equation* for the velocity from part (a), determine the velocity at a time somewhere in the middle of the motion (e.g., at 1.5 seconds). How does this value compare to the value in the data column? Are they similar? Would you expect them to be *exactly* the same?

**c.** Next, repeat the above process for acceleration. Change the plot so you are viewing an acceleration-time graph and use the appropriate parameter from the best-fit *position* equation to model the acceleration graph.

**Note**: Your software might not give you an option to model using a constant. If this is the case, you can use a linear function with a slope of zero. Print out or sketch a copy of your graph and model below.

 

**d.** Finally, you might have noticed that while the position model looks pretty good over the entire data set, the velocity and acceleration models may not look very good at the ends. Do you think this is a problem with the model or the data? Explain briefly. **Hint**: Think about what we learned about how the software *calculates* the velocity and acceleration data at the end of Unit 3.

---

As shown in this activity, the position graph contains all the information needed to obtain both the velocity graph and the acceleration graph. However, the converse is not true. For example, an acceleration graph does not tell us the initial velocity or the initial position. Thus, there are many (infinitely many) velocity graphs that are consistent with a particular acceleration graph. Similarly, there are many position graphs that are consistent with a particular velocity graph.

## 4.6   SOLVING PROBLEMS WITH CONSTANT ACCELERATION

As we have seen, if an object moves with a constant acceleration we can determine its velocity and position as functions of time using the kinematic equations (rewritten here for convenience):

$$v_x(t) = v_{0x} + a_x t$$

$$x(t) = x_0 + v_{0x}t + \frac{1}{2}a_x t^2$$

where $a_x$ is the (constant) acceleration, $v_{0x}$ the initial velocity, $x_0$ the initial position, and we have assumed the initial time is zero ($t_0 = 0$). It is frequently the case that objects move with a constant (or nearly-constant) acceleration, and the kinematic equations allow us to mathematically determine the motion for a wide variety of situations. In this section we demonstrate how to solve certain types of problems using these two kinematic equations. Regardless of the specific situation, the following steps are helpful as a general approach to solving kinematic problems:

Step 1. Think carefully about the problem and sketch the basic situation, including a coordinate system.

Step 2. Write down what you know and what you're trying to find and then sketch any diagrams or graphs that might be helpful.

Step 3. Write down the relevant equations and determine which one(s) will be most useful.

Step 4. Carry out the calculations and double-check that your answers are reasonable.

---

### 4.6.1. Activity: A Straightforward Situation

Imagine you are driving your car onto the freeway. You are originally traveling at a speed of 13.0 m/s (approximately 30 mph) when you push down on the accelerator, causing the car to accelerate at a (constant) rate of 2.70 m/s$^2$. We want to determine how fast the car will be traveling after 5 seconds and how far it will travel during this time.

**a.** Think through the problem and define an appropriate coordinate system (Step 1). Discuss this with your group and then draw a rough sketch of the situation that includes your coordinate system (be sure to specify the location of the origin and the positive direction).

**b.** Making use of the above coordinate system, write down any information you know (initial position, initial velocity, etc.), any information you don't know, and what are you ultimately trying to find (Step 2).

**c.** Based on what we learned in the previous sections, make a rough sketch of what you expect the velocity-time graph to look like (Step 2). Include as much quantitative information on your graph as you can (though there may still be some unknowns on the graph).

**d.** Write down the two kinematic equations and explain which one would be most useful for finding the speed of the car after the 5-second interval (Step 3). Which one would be most useful for finding the distance traveled during this time?

**e.** If you have all the information you need, calculate the speed of the car after the 5-second interval (Step 4). Indicate this point on your graph in part (c). Give you answer in both meters per second and miles per hour. Does your answer seem reasonable? Explain briefly.

**f.** Finally, calculate how *far* the car travels during this time interval (Step 4). Does your answer seem reasonable? Explain briefly. (Note that we are asking how far the car traveled during the 5 seconds in which it is accelerating; in other words, how much did the car's position *change* during this time?)

The above problem is relatively simple, but it provides a good example for demonstrating the problem-solving process. Although you might be tempted to skip some steps, we advise you to go through each step for every problem. These steps help you to think carefully about the problem and to organize the information in a manner that promotes finding a solution.

### 4.6.2. Activity: Another Straightforward Situation

Microgen/Adobe Stock

An Olympic diver jumps off the high platform that is 10.0 m above the water. They jump with an upward speed of 1.50 m/s, and it takes them 1.59 s to reach the water's surface. Our goal is to determine the diver's *acceleration* during their fall (assuming it's constant) and their speed when they reach the water's surface. **Note:** You are supposed to *calculate* the acceleration here, not just cite it from memory.

**a.** Although this problem is technically two-dimensional, we will assume that the motion occurs only in the vertical direction. Discuss this problem with your group and then draw a rough sketch of the situation that includes your coordinate system (be sure to specify the location of the origin and the positive direction).

**b.** Making use of the above coordinate system, write down any information you know (initial position, initial velocity, etc.), any information you don't know, and what are you ultimately trying to find. Be mindful of signs.

**c.** Make a rough sketch of both a velocity-time graph and a position-time graph. Include as much quantitative information on these graphs as you can (though there may still be some unknowns on the graphs). **Note**: Your position-time graph need not be quantitatively accurate, just try to get the correct shape and then label any points you know.

**d.** Write down the two kinematic equations and explain which one would be most useful for finding the acceleration and which one would be most useful for finding the speed at the water's surface.

**e.** If you have all the information you need, calculate the diver's acceleration. Does your answer seem reasonable? Explain briefly.

**f.** Finally, calculate the diver's speed at the water's surface. Does your answer seem reasonable? Indicate this point on your graph in part (c).

It is worth commenting on the choice of coordinate system in this problem. The motion in this case is in the vertical direction, and you may have chosen to represent your position using the variable $y$ instead of $x$. The choice is arbitrary, and for one-dimensional motion, you can choose whichever coordinate you wish (though if you choose $y$, be sure all the variables in the kinematic equations refer to the $y$-direction: $v_y$, $a_y$, etc.).

More important is the direction of the positive axis and the location of the origin. We are only told that the diving platform is 10 m above the water, so there are multiple options to choose from. You should align your coordinate axis along the direction of motion (vertically), but whether the positive direction points up or down is your choice. Furthermore, although you are free to select the location of the origin anywhere along the axis, there are two options that make the most sense based on this problem: the surface of the water or the location of the diving platform.

While your choices will affect the values of the parameters in the problem, barring any mathematical errors along the way, you will arrive at the correct answer no matter what you choose; namely, an acceleration of approximately 9.8 m/s$^2$ directed downward toward the water's surface. But note that the *sign* of your result will depend on whether you choose the positive direction as up or down, so it's essential to clearly define a coordinate system!

### 4.6.3. Activity: A More Complicated Situation

Imagine you are riding your bike along a campus sidewalk at a speed of 6.00 m/s. All of a sudden, a door from one of the buildings opens and a large cart is rolled out onto the sidewalk 5.50 m in front of you. You immediately apply the brakes on your bike, resulting in your speed decreasing at a constant rate of 3.60 m/s$^2$. Do you manage to stop before running into the cart?

**a.** As before, begin by discussing this problem with your partners and then make a rough sketch that includes a coordinate system.

**b.** Making use of the above coordinate system, write down any information you know (initial position, initial velocity, etc.), any information you don't know, and what are you ultimately trying to find. As always, be mindful of signs.

**c.** Make a rough sketch of both a velocity-time graph and a position-time graph, including as much quantitative information on these graphs as you can (though there may still be some unknowns on the graph). **Note**: Your position-time graph need not be quantitatively accurate, just try to get the correct shape and then label any points (positions and times) you know.

**d.** Write down the two kinematic equations. We ultimately want to know whether the bike crashes into the cart, so we are looking to find the final position of the bike. Unfortunately, we can't use the position equation directly to get the answer. Discuss with your group how you might solve this problem using these two kinematic equations. Briefly describe your plan of action and carry out your procedure, keeping your work as clear as possible. Does your final answer seem reasonable? Explain briefly.

One way to solve the previous problem is to first use the velocity equation to find the time it takes to stop, and then use this time in the position equation to determine the final position. If you travel less than 5.50 m (your original distance from the cart), then you were able to stop in time.[4] But notice that we don't really care how long it takes to stop, so the time found from the velocity equation is simply an intermediate step needed to link the velocity and position equations. It turns out there's a way to combine these two equations so that the time variable is eliminated, producing a new equation that links only the position, velocity, and acceleration. We derive this equation in this next activity.

**The Third Kinematic Equation**

We start with the position and velocity kinematic equations, using the general case where the initial time $t_0$ is not necessarily zero:

$$v_x(t) = v_{0x} + a_x \Delta t$$

$$x(t) = x_0 + v_{0x} \Delta t + \frac{1}{2} a_x (\Delta t)^2$$

Note these equations are written in terms of the *change* in time $\Delta t = t - t_0$.

---

[4] Another approach is to use the position equation to calculate how much time it takes to travel 5.50 m and then use this time in the velocity equation to determine if you are still moving forward at that instant. The downside of this approach is that if you *do* stop in time, you will never reach 5.50 m from where you started. In this case, the position equation will have no solution (more accurately, it will have no *real* solution).

### 4.6.4. Activity: Deriving a Third Kinematic Equation

**a.** Begin by re-writing the two equations above in terms of $\Delta x = x(t) - x_0$ and $\Delta v_x = v_x(t) - v_{0x}$ (this is straightforward—don't overanalyze!).

**b.** Next, solve the new velocity equation for $\Delta t$ and substitute this into the new position equation. Then simplify and show that you get $2a_x \Delta x = 2v_{0x}\Delta v_x + (\Delta v_x)^2$.

**c.** Finally, we want to make use of the fact that $\Delta v_x = v_x(t) - v_{0x}$. However, recall that our goal is to eliminate time as a variable, so it doesn't make much sense to continue to write the velocity as a function of time. With this in mind, replace $\Delta v_x$ with $v_x - v_{0x}$ in your result from part (b) and show that you end up with the following expression (it helps to keep your work neat and organized, as there's a bit of algebra here):

$$v_x{}^2 = v_{0x}{}^2 + 2a_x \Delta x \qquad (4.4)$$

In reading this equation, $v_{0x}{}^2$ can be considered to be the velocity (squared) at the initial *position*, while $v_x{}^2$ is the velocity (squared) at the final position (after the position has changed by $\Delta x$).[5] Note that this kinematic equation achieves our goal of linking the position, velocity, and acceleration without an explicit reference to time. (Once again, keep in mind that this equation, just like the others, is only valid when the acceleration is *constant*.)

---

[5] You might think that because $v_x$ is now technically written as a function of position, we should indicate this dependence by writing $v_x(x)$. While this does seem reasonable, it turns out that such notation is seldom used.

**4.6.5.  Activity: A Slightly More Complicated Situation Revisited**

Now that we've eliminated time from the kinematic equations, we can use the new equation to simplify the solution when the time is not needed. As an example, try re-solving the problem from Activity 4.6.3 using Eq. (4.4). Do you get the same result as before (you should)? Here is the problem statement again for convenience: Imagine you are riding your bike along a campus sidewalk at a speed of 6.00 m/s. All of a sudden, a door from one of the buildings opens and a large cart is rolled out onto the sidewalk 5.50 m in front of you. You immediately apply the brakes on your bike, resulting in your speed decreasing at a constant rate of 3.60 m/s². Do you manage to stop before running into the cart?

### 4.7  PROBLEM SOLVING

**Fig. 4.3.** Apollo 16 astronaut John Young jumping on the moon. Image credit: NASA.

During the Apollo 16 mission of 1972, astronauts spent 11 days on the Moon conducting scientific experiments. Among the video clips transmitted back to Earth during the mission is astronaut John Young jumping on the moon (see Fig. 4.3). In this first problem, you will analyze the motion of Young's jump to determine his acceleration (due to the gravitational pull of the moon). Depending on the video analysis software being used, your instructor may have specific instructions on how to perform certain aspects of the analysis.

#### 4.7.1. Activity: Jumping on the Moon

a. **Watch the Movie:** Begin by watching the movie "Moon Jump" to get a sense of what happens (and to appreciate this moment in history).

b. **Set the Scale and Origin:** Now advance the movie to the frame just before the astronaut jumps for the first time and place the origin on the ground at the bottom of his feet. Then scale the movie using the flagpole, which has a known height of 2.2 m (be sure to select the entire flag pole, as the bottom portion can be hard to see).

c. **Record Position Data:** Before taking any data, advance the movie frame-by-frame until you are certain the astronaut is in the air (it's a little hard to tell because a bunch of moon dust is kicked up when he jumps—just do your best). Now find a spot on the astronaut that you can track throughout the jump, such as the top of the backpack, and click through the video, doing your best to keep the cursor target lined up with the same point on the astronaut's backpack in each frame. Be sure to stop taking data *before* the astronaut reaches the ground. Tracking this motion is not simple, particularly at the top of the trajectory; just do the best you can! You should have about 40 data points (or slightly fewer) at the end. **Note:** If you really mess up, you can always start over.

d. **Fit the Data:** We are only interested in the *vertical position* data (typically the *y*-position), so make sure this is the only quantity being plotted. Then use the "Curve Fit" feature to fit the position-time data (choose an appropriate function for the fit, and enable "Time Offset" to account for the fact that the astronaut does not jump at time $t = 0$). Be sure to fit only the portion of the data where the astronaut is in the air (if you were careful when selecting your start and end points, then all of your data will be used). Once you have the fit, write down the equation governing the astronaut's motion, being sure to include all appropriate units.

e. **Determine the Acceleration on the Moon:** Use the appropriate curve-fit parameter to determine the (vertical component) of the astronaut's acceleration. Be sure to show how you obtained your answer.

f. **Interpret the Other Coefficients:** Similarly, determine the astronaut's initial velocity in the vertical direction and initial (vertical) position of the point on the spacesuit you chose. Do these values make sense (where is your origin)? Explain briefly.

g. **Moon Versus Earth:** Compare the value you obtained for the astronaut's vertical acceleration to the theoretical value of $-1.6\,\text{m/s}^2$ on the Moon. How does this compare to the acceleration you measured on Earth in Activity 4.3.3?

---

The next problem deals with the kinematic equations in a more complicated situation. We are interested in determining the appropriate speed limit for cars driving through a school zone. To do so, we will need a few pieces of information:

- For pedestrian safety, experts suggest that a vehicle should be able to come to a complete stop within a distance of 8.0 m from the instant the driver first sees someone walk into the road.
- Hard breaking in a typical car causes the vehicle to slow down at a rate of $5.0\,\text{m/s}^2$.
- There is some reaction time from the instant a driver sees a child until their foot pushes on the brake pedal, and this needs to be considered. A good estimate for the reaction time of an attentive driver is about 0.5 s.

### 4.7.2. Activity: Speed Limit Designation

a. Use the information above to find the maximum speed from which a driver can safely stop should a child run into the road. Convert your final answer into miles per hour and round in the appropriate manner to determine what the speed limit sign should read.

　　You will need to think carefully about how to solve this problem (we recommend following the problem-solving procedure outlined in Section 4.6). Remember, there are two *separate* portions of motion:
- The time *before* the brakes are applied, when the driver is reacting and the car is traveling at a constant speed.

- The time during the actual braking when the car is slowing down at a constant rate.

These two parts are linked by what is known as a "constraint." In this case, we want the *total* distance traveled (including both parts of the motion) to be a maximum of 8.0 m for a car that is initially traveling at the unknown speed limit. **Hint**: For the two parts of the motion, write down kinematic equations that give the distance traveled during each portion in terms of the (unknown) initial speed. Then write down a third equation that describes the constraint. These three equations allow you to solve for the initial speed.

**b.** What *percentage* of the stopping distance of 8.0 m is necessary just to account for the driver's reaction time? (Can you see why distracted driving can be so dangerous?)

# UNIT 5: FORCE, MASS, AND MOTION

*NASA*

*The Saturn V launch vehicle carrying astronauts Neil A. Armstrong, Michael Collins, and Edwin "Buzz" E. Aldrin, Jr., lifted off on July 16, 1969, as part of the Apollo 11 mission to land a human on the moon. The rocket generated a thrust of over seven million pounds during liftoff, lumbering off the launch pad slowly at first but then picking up speed. In fact, its speed increased ever more rapidly as it continued upward, even though the ejected fuel created a constant upward force on the rocket. As you study the fundamental relationships between force, mass, and motion in this unit, you should be able to determine why the rocket moved the way it did.*

# UNIT 5: FORCE, MASS, AND MOTION

## OBJECTIVES

1. To devise a method for applying constant forces to objects.

2. To find the mathematical relationship between force and motion and understand how multiple forces can be combined.

3. To develop an understanding of the concept of inertia and mass.

4. To formulate statements of *Newton's First and Second Laws of Motion* for one-dimensional motion.

## 5.1 OVERVIEW

In our study thus far we have learned to observe and describe motion using words, graphs, and equations. The next step is to investigate the *causes* of motion. The motion of an object is obviously influenced by pushes or pulls, and these can arise from direct contact, such as from your hand or the ground, or without physical contact, such as by gravity or electromagnetism.

Casual observations indicate that how an object responds to a push or pull depends on the "amount of stuff" the object contains—it's easier to push a shopping cart than a truck. In physics, we usually refer to a push or pull as a *force* and the "amount of stuff" as the *mass*. In this unit we will explore intuitive ideas about force and mass and then measure how a force affects the motion of an object. Finally, based on what we learn, we will formulate two laws of motion developed by Isaac Newton in the seventeenth century.

Newton's laws of motion are very powerful. When the forces on a system are known, Newton's laws can be used to describe and predict its behavior. This predictive ability is of tremendous importance for designing bridges that don't collapse or cars that stop reliably when the brake pedal is pushed. Moreover, an understanding of these laws of motion allows scientists to deduce the nature of both interplanetary and intermolecular forces simply based on observations of motions. As stated by Newton himself,

… the phenomena of motions [can be used] to investigate the forces of Nature, and then … these forces [can be used] to demonstrate other phenomena … the motions of the planets, the comets, the moon and the sea.

**Note**: The classical laws of motion we will develop in this Activity Guide provide, for all practical purposes, "exact" descriptions of the motions of everyday objects traveling at ordinary speeds. However, these laws of motion are not complete if describing items like subatomic particles or black holes. Describing such objects requires the use of quantum theory and/or relativity.

## FORCE AND MOTION

### 5.2   FORCE WITHOUT MOTION

In our previous study of motion, we focused our attention on describing the motion rather than understanding its causes. Yet from everyday experience we know that force and motion are related. For example, to get a bicycle moving you must push down on the pedals. But once moving, do you need to continue pushing on the pedals to keep the bicycle moving in a steady manner? Does it depend on whether you are on the street or the grass? And what if you want to change the bicycle's motion in some way?

Before we can study the relationship between force and motion, we must devise a useful definition of force and develop reliable ways to measure it. Only then can we investigate how forces affect motion. We will begin simply by considering how to apply a *fixed* amount of force to an object.

The activities in this section are somewhat open-ended. Although the following items should be available, you might not need them all:

- 1 table clamp
- 1 rod
- 1 ruler
- Multiple rubber bands, #14
- 2 mass pans, 50 g
- 2 masses, 50 g
- 5 masses, 100 g

#### What Is Force?

What is force and how is it measured? The word *force* is a very common part of everyday language. One of the major tasks in this unit is to help you move from an informal understanding of the term force as a push or a pull, to a more precise, quantitative definition that will be useful in relating force to motion.

---

#### 5.2.1.  Activity: Simple Ideas About Force

**a.**  List some examples of forces. How would you define the word *force* in your own words?

**b.**  Can you think of a way to *measure* how large a given force is? Does force have a direction? Explain whether you think force might be a vector quantity.

---

**Defining a Unit of Force**

To explore the relationship between force and motion, we need to determine how to apply different, reproducible forces to an object. There are many ways to do this; we will make use of rubber bands.

---

**5.2.2. Activity: Applying a Rubber Band Force**

a. Explain how you can apply a small force to an object using a rubber band (for now, we will assume the object is fixed in place so that it doesn't move). How would you apply a larger force to this object? Does the force you apply with a rubber band have a direction? Explain briefly.

b. How can you apply a force with a changing magnitude? How about a force with a changing direction?

c. To *quantify* force, we need to define a specific measure of what it means to pull with a certain strength. While such a definition is arbitrary, it's helpful if it can be explained easily to others. Discuss this with your group and come up with a definition using your rubber band; in other words, finish the sentence, "To apply one unit of force (call it whatever you want) to an object, you must ... "

d. Using the definition above, how might you apply 2, 3, or 10 units of force? Can you do this with a single rubber band? Is there a limit to how much force you can apply with a single rubber band? If so, explain how you can use multiple rubber bands to apply 2, 3, or 10 units of force.

e. How might you apply a force that's *one-half* or *one-third* as large as your defined unit of force? Explain briefly.

---

Pulling on an object using a rubber band is an easy way to apply a force to an object; the more the rubber band is stretched, the larger the applied force (Fig. 5.1). In addition, the object is pulled in a direction *aligned* with the rubber band, and it's easy to change the direction of the force by simply pulling the rubber band in a different direction. Having both magnitude and direction suggest that force is a vector quantity, something most people find quite intuitive. As we will see, one can show that forces can be added together even if they are pointing in different directions. The result, which we state here without proof, is that forces add *vectorially* (like vectors).

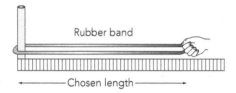

Fig. 5.1. Pulling a rubber band out to a chosen length.

Once we have defined a unit of force (e.g., 4 cm of stretch of a particular type of rubber band), it might seem reasonable to assume that doubling the amount of stretch will double the amount of force. This turns out to be true if the amount of stretch is not too large. When the amount of force is proportional to the amount of stretch, we say the force obeys *Hooke's* law, named after the seventeenth century British physicist Robert Hooke. Springs typically obey Hooke's law quite well, and it is common to use springs in simple force sensors.

However, if the stretch gets too large the force will start to vary *nonlinearly* with the amount of stretch. In this case, doubling the amount of stretch will *not* double the force. Moreover, there is clearly a limit to how much force can be applied with a single rubber band because it will eventually break! But we can always use two (or more) identical rubber bands to apply larger and larger forces. In fact, by using various combinations of rubber bands stretched by different amounts, we can, at least in principle, apply any amount of force we want.

## 5.3 MOTION FROM A CONSTANT FORCE

Now that we have a method for exerting forces of varying magnitude, we can try applying a constant force to an object that can move. This turns out to be harder than it sounds because it's challenging to maintain a fixed amount of stretch while moving. We will begin by pulling a person across the room using a constant force in two different situations: (1) when sliding across the floor on a burlap sack (or a blanket), and (2) when riding on a low-friction cart.

By tracking the motion of the person using a motion sensor, we will get our first look at how force and motion are related. To do the activities in this section you will need the following items:

- 1 m stick
- Several bungee cords or a large spring scale
- 1 data-acquisition system

- 1 motion sensor
- 1 large burlap sack (or blanket)
- 1 kinesthetic cart (or large skateboard)

### Predicting Motion from a Constant Pull

Consider pulling someone across a level floor while maintaining a constant force as a motion sensor tracks their position. You first pull the person when they are sitting on a large burlap sack. You then pull them while they are sitting on a rolling cart or skateboard. These two situations are shown in Fig. 5.2.

(a)  (b)

**Fig. 5.2.** Being pulled with a constant force under two circumstances: (a) *sliding* along a smooth level floor while sitting on a burlap sack and (b) *rolling* along a level floor on a low-friction cart.

---

#### 5.3.1. Activity: Predicting the Velocity of a Person Being Pulled with a Constant Force

**a.** Consider the situation in Fig. 5.2(a). What do you think the *velocity-time* graph will look like if the person starts from rest and *slides* along the floor while being pulled away from a motion sensor with a *constant* applied force? Sketch the shape of your *predicted* graph below.

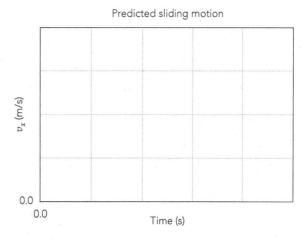

**b.** Now consider the situation in Fig. 5.2(b). What do you think the *velocity-time* graph will look like if the person starts from rest and *rolls* along the floor while being pulled away from a motion sensor with a *constant* applied force? Sketch the shape of your *predicted* graph below.

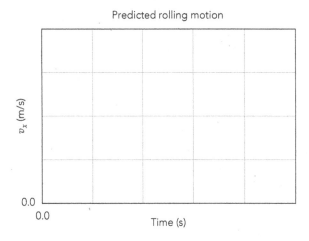

**c.** Do your predicted graphs look similar or different? Briefly explain why.

### Observing Motion from a Constant Pull

This experiment is best done as a class demonstration. Work with the rest of the class to create and record data for both sliding and rolling motions when pulling with a constant force while tracking the position using a motion sensor. You can use some bungee cords (or a large spring scale) stretched out to a fixed distance to create a constant pulling force. If the person being pulled holds a meter stick along the bungee cords, the person doing the pulling can try to keep the amount of stretch fairly constant (although this is not easy). The person should be pulled away from the motion sensor with plenty of space before encountering any obstacles. These experiments are challenging to perform, so you might need to try them a few times to get clean data.

#### 5.3.2. Activity: Observing the Velocity of a Person Being Pulled with a Constant Force

**a.** Carry out the *sliding* experiment and sketch the velocity-time graph for this situation below. **Note**: The experimental data for this experiment might be a little "noisy"; we are mainly interested in the overall shape of the graph and not the minor bumps associated with wobbles in the spring or bungee cord.

Observed sliding motion

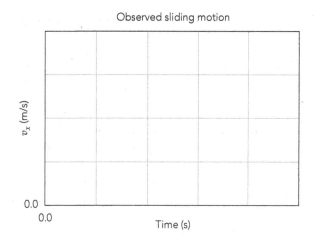

b. Examine the overall shape of the velocity-time graph for *sliding* motion. After the motion has started, does the velocity appear to be constant or changing? What does this graph tell you about the acceleration? Explain briefly.

c. Next, carry out the *rolling* experiment and sketch the velocity-time graph for this situation. (As with the previous experiment, you should ignore the minor bumps and focus on the overall trend of the graph.)

Observed rolling motion

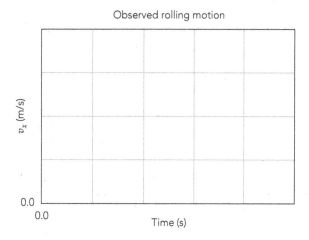

d. Examine the overall shape of the velocity-time graph for the *rolling* motion. Does the velocity appear to be constant or changing? What does this graph tell you about the acceleration? Explain briefly.

**e.** How did your observations in these experiments compare to your earlier predictions? Did either result surprise you?

**f.** Based on your answers to parts (b) and (d), describe the *acceleration* that results from a constant applied force for both sliding motion (when friction is significant) and for rolling motion (when friction is small). Again, we are interested in the acceleration after the motion has been established. Make a sketch of the *acceleration-time* graph for both situations on the axes below, being sure to label which curve corresponds to rolling and which to sliding.

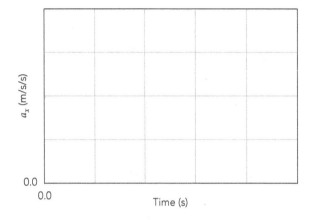

## Rolling versus Sliding Motions

When an object is *sliding*, you probably observed that a constant force leads to a roughly constant velocity (once the motion has been established). On the other hand, a *rolling* object appears to move with a constant (non-zero) acceleration under the influence of a constant force (its velocity increases linearly with time). The sliding motion involves a significant amount of friction while the rolling motion does not, and this difference leads to the different types of motion.[1]

Ultimately, we want to understand both types of motion, but as is often the case in physics we begin with the simplest situation possible. Because our goal in this unit is to determine how forces affect motion, we will start with situations in which friction can be ignored (like the rolling situation). By doing so, we attempt to focus solely on the applied force and the resulting motion, without the added complication of friction. Once we have determined the relationship between force and motion, we will return to the question of how friction can be incorporated into the laws of motion.

---

[1] If you already know some physics, you might be somewhat perplexed as to how the pulling force and the friction force end up being perfectly balanced. This puzzle is discussed in an article by R. Morrow, A. Grant, and D.P. Jackson, "A strange behavior of friction," *Phys. Teach.* **37**, 412–415 (1999).

**A word of caution:** No matter how hard we try, it is impossible to eliminate friction entirely. Thus, if you look closely at the data in the following experiments you will likely see subtle signs that friction is present. For now, we ask that you simply look past these small effects and focus your attention on the main features of the experiments.

## 5.4   STANDARD UNITS FOR MEASURING FORCE

So far, we have been measuring forces using arbitrary units related to the amount of stretch of a particular rubber band. But if you want to communicate force measurements with others, it would obviously be convenient if everyone agreed to use the same unit of force. There are several force units in common use, two of which are the *pound* (abbreviated lb) and the *newton* (abbreviated N). You are probably more familiar with the pound, but the newton is more common in science and is what we will use throughout this Activity Guide. For reference, one pound is equivalent to (approximately) 4.45 newtons. Precisely how (and why) these units are defined the way they are is not important; all that matters is that they are accepted standards, just like an inch, a meter, or a mile.

A common device for measuring forces consists of a spring with a scale attached to it that is marked in newtons. These *spring scales* are very similar to using a rubber band but have been *calibrated* so that the markings correspond to the newton force standard. A less common but very useful way to measure force is an electronic force sensor attached to a computer data-acquisition system.

In the activities in this section we will measure forces with both spring scales and electronic force sensors using the following equipment:

- 1 ruler
- 6 rubber bands, #14
- 1 spring scale, 10 N (or 20 N)
- 1 data-acquisition system
- 1 electronic force sensor

### Measuring Force Using a Spring Scale

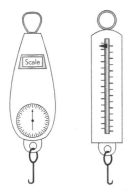

Fig. 5.3. Two types of spring scales used to measure force.

Begin by holding the spring scale with nothing pulling on it. Look carefully at the reading on the scale. Does it read zero? Typically, a spring scale is calibrated so that it reads zero when hanging vertically with the hook at the bottom (as in Fig. 5.3). If you try to use a spring scale when oriented horizontally, such as when sitting on a table, you will see that it no longer reads zero when nothing is pulling on it (try it and see). In general, a force sensor is designed to be used in a specific orientation, and if you change the orientation there will be a systematic error that will render your measurements inaccurate (unless properly accounted for). Most force sensors are designed so the zero point can be reset as needed. As we will see, resetting the zero point on an electronic force sensor is quite simple (the *tare* button on an electronic scale is an example of resetting the zero point). The bottom line is that the orientation of a force sensor matters, so you should always zero an electronic force sensor in the appropriate orientation *before* making any measurements.

### 5.4.1. Activity: Calibrating Your Rubber Band Force

**a.** Begin by holding the spring scale with nothing pulling on it and make sure that it reads zero. Now try pulling on the spring scale with a rubber band. As you pull harder, both the rubber band and the spring become more stretched. Try pulling on the force sensor using the rubber band force unit you defined in Activity 5.2.2. How many newtons does the spring scale read? How many newtons does the spring scale read if you pull with two, three, or four rubber band units? (If your spring scale maxes out, you'll need to get one with a larger range.)

**b.** Using the measurements you just made, determine a conversion factor that can be used to convert your rubber band units into newtons.

Although spring scales are very convenient, they have some disadvantages. For example, because it's not simple to reset the zero point, spring scales can only be used in one orientation while remaining quantitatively accurate. Another issue is that there's no way to measure a *push* using a spring scale. For these reasons, it is usually more convenient to use an electronic force sensor. An added bonus is that an electronic sensor allows us to graph our measurements in real-time.

### Measuring Force Using an Electronic Force Sensor

Just like a spring scale, an electronic force sensor needs to be both calibrated and zeroed to obtain accurate readings. Fortunately, our force sensors are pre-calibrated, and while there may be times when a force sensor needs to be re-calibrated, the process is straightforward (your instructor can help you with this procedure if necessary). An electronic force sensor can be used in any orientation, but it is important to make sure the force sensor is *zeroed before each measurement*, particularly when the orientation of the sensor has changed. The following activity should make this clear.

### 5.4.2. Activity: Experimenting with a Force Sensor

**a.** To begin, make sure the force sensor is set to the ±10-N scale (there is often a switch to set the range directly on the sensor), and set up an experiment that runs for about 10 seconds. For this experiment there should be nothing attached to the force sensor, and it should be zeroed with the hook pointing down. Once the experiment begins, hold the sensor steady with the hook pointing down for 2 or 3 seconds, and then change the orientation so the hook is pointing up and hold it steady for 2 or 3 seconds. When the experiment ends, re-scale the graph so the range goes from $-1$ to $+1$ newton. Comment on what you observe and explain what you think is happening.

 

    **b.** As you should have noticed, the *orientation* of the sensor affects the reading. If you zero the sensor with the hook pointing down, then the sensor will only read zero when the hook points down. Try the above experiment again, but this time zero the sensor with the hook pointing up before starting the experiment. What do you observe this time?

 

    **c.** For the next experiment, you are going to *gently* pull and push on the hook (with a maximum force less than 10 N). Begin by zeroing the force sensor in a specific orientation and then maintain this orientation throughout the experiment. Comment on what you observe in the force-time graph as you gently pull and push on the sensor.

 

    **d.** Finally, begin a new experiment (don't forget to zero the sensor first), and this time pull on the hook harder than before. Continue pulling harder until the reading flattens out. Then try pushing on the hook until the reading flattens out again. Based on your observations, describe the effective range of the force sensor with its current settings.

---

The previous experiment demonstrates the usefulness of an electronic force sensor: you can obtain a real-time graphical display of the force being exerted on the sensor; the sensor can be used in any orientation (as long as it is properly zeroed); and force sensor readings can be positive *or* negative, with positive forces corresponding to pulls and negative forces corresponding to pushes.

We already discussed that force is a vector quantity, and like all vectors it consists of a magnitude and a direction. Just like position, velocity, and acceleration, a force will, in general, have more than one component: $\vec{F} = F_x\hat{x} + F_y\hat{y} + F_z\hat{z}$. However, if we consider forces that act along a single axis (*one-dimensional forces*), then the force vector will only have one component. Assuming the force is in the $x$-direction, the force vector will then be given by $\vec{F} = F_x \hat{x}$. Just as we did for motion, we can neglect the unit vector $\hat{x}$ and express the force simply as $F_x$.

Like all vector components, $F_x$ can be positive or negative, and in one dimension we can determine the direction of the force simply by looking at the sign of this component. Like the motion sensor, we use a standard convention for the force sensor: the direction away from the force sensor (pulling on the sensor) defines the positive direction. This convention is arbitrary, of course, and can be overridden using the software, but we will seldom have a need to do so.

## 5.5  RELATING ACCELERATION AND FORCE

In Section 4.3 we observed that a fan pushing on a cart with a *constant* thrust caused it to accelerate at a *constant* rate. Now that we have a more thorough understanding of how forces are defined and measured, we are ready to quantitatively investigate the relationship between the applied force and the resulting acceleration. You may find it helpful to review the results of the activities in Section 4.3.

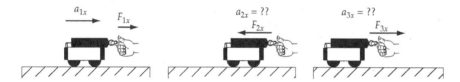

**Fig. 5.4.** How does the measured acceleration along an axis for a low-friction cart change when the applied force on it is changed?

---

**5.5.1. Activity: Predicting Acceleration**

**a.** In Section 4.3, we used a fan to provide a constant thrust on a cart and observed that the cart moved with a constant acceleration. We'll assume that the fan thrust acted like a force, such as we have seen with a pull of a rubber band. When the force was directed in the positive direction (pushing away from the motion sensor), was the acceleration positive or negative?

**b.** When the force was directed in the negative direction (pushing toward the motion sensor), was the acceleration positive or negative?

**c.** Did a larger magnitude force result in a larger magnitude acceleration?

**d.** Now suppose you push and pull on a force sensor attached firmly to a low-friction cart so that the cart moves in response to your force (as in Fig. 5.4). Assume you obtain a force-time graph as shown below. In light of your answers to parts (a), (b), and (c) above, predict what the acceleration of the cart will look like while moving in response to this force.

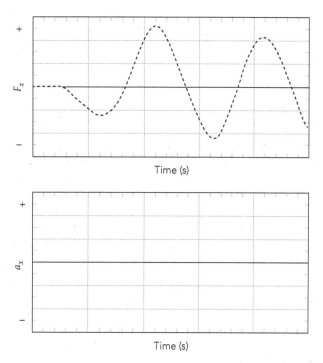

Now that you've made a specific prediction, it's time to try the experiment. To investigate how acceleration and force are related, you will push and pull on a force sensor that is attached to a cart and record the motion of the cart using a motion sensor. To accomplish this you will need:

- 1 low-friction dynamics cart
- 2 mass bars, 500 g (to add mass to the cart)
- 1 smooth cart track or level surface 1–3 m long
- 1 data-acquisition system
- 1 ultrasonic motion sensor
- 1 force sensor (with a hook on its sensitive end)
- 1 adapter bracket (to attach a force sensor to the cart, if needed)
- 1 spring scale, 10 N, or hanging masses (to calibrate the force sensor, if needed)

In the following activity you will measure both the force and motion simultaneously (see Fig. 5.5). To accomplish this, attach a force sensor firmly to a cart and make sure the force sensor is on the ±10 N scale. Put a motion sensor at one end of the track and set up the software to display a force-time graph and an acceleration-time graph (one above the other) with an experiment length of about 5 seconds.

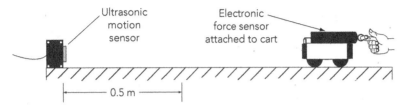

**Fig. 5.5.** Setup showing a motion sensor tracking the acceleration of a cart rolling on a level track as a force sensor detects the pushes and pulls on it. Be sure to start the car at least 0.5 m away from the motion detector so that you do not get too close.

### 5.5.2. Activity: Measuring Force and Acceleration

**a.** Before performing this experiment, be sure to zero the force sensor with no push/pull. Then, record data as you firmly hold the hook on the end of the force sensor and *smoothly* roll the cart back and forth along the track (try to apply a force that is aligned with the track). Repeat this process until you have smooth, reliable data. Make a sketch of your graphs below and comment on what you observe. **Note:** You should observe an *obvious* similarity between the two graphs; if you're unsure, ask for help!

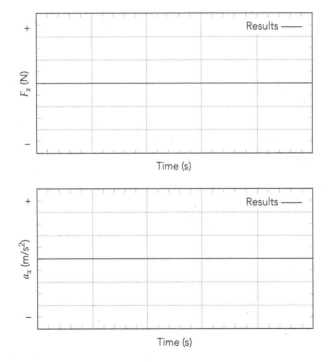

> **b.** Describe how the force and acceleration appear to be related mathematically.

_____

### The Relationship Between Force and Acceleration

The fact that the force and acceleration graphs have exactly the same shape suggests that force and acceleration are related to each other in some simple way. When faced with such a situation, one useful way to proceed is to plot one variable versus the other. Fortunately, because we have already taken data for both force and acceleration, it is a relatively simple matter to make a plot of force versus acceleration.

_____

#### 5.5.3. Activity: The Mathematical Relationship Between Force and Acceleration

> **a.** Use the software to create a graph of force versus acceleration using the data you just took. The easiest way to do this is to change the time on your force-time graph to acceleration.[2] **Note**: Before proceeding, make sure that your force-time graph is being plotted as a "scatter plot," in which the data points are plotted _without_ connecting the points together (ask your instructor how to accomplish this if needed). Connecting the points is useful when plotting force versus time, but it makes things more confusing when plotting force versus acceleration. Print out or sketch the graph, making sure to label the axes.

> **b.** The data in this experiment should look like it lies on a straight line that passes through the origin, indicating that force and acceleration are _proportional to each other_. (As you know, all measurements have some inherent uncertainty so the data will not form a _perfect_ line.) Write down an equation that relates the force $F_x$ and the acceleration $a_x$, using the symbol $m$ to represent the proportionality constant (i.e., the _slope_ of the line).

_____

[2] It is conventional to plot the independent variable on the $x$-axis and the dependent variable on the $y$-axis. Following this convention, we _should_ plot acceleration on the $y$-axis and force on the $x$-axis. However, in this particular case it is more convenient to plot it the other way.

**c.** Using the computer, perform a proportional (or linear) fit to the data to determine the specific value for the slope in your experiment (don't forget units). Write down the equation for the fit below. Can you speculate as to what the slope represents physically?

You have just discovered a law of proportionality between an applied force and an object's acceleration for a one-dimensional situation in which there is almost no friction. Mathematically, we can write this as

$$F_x = ma_x$$

where $m$ is a constant (the proportionality constant in the above experiment) that depends on the object. From the perspective of the cart's motion, if you apply a force $F_x$ to an object, it will respond with an acceleration

$$a_x = \frac{F_x}{m}$$

This relationship is almost, but not quite, one of Newton's famous laws of motion. However, before we can finalize this law we need to explore how the properties of the object being accelerated affect the proportionality constant. In addition, we also need to explore what happens when more than one force acts on an object at the same time.

## THE NET FORCE AND NEWTON'S LAWS

### 5.6   THE NET FORCE: ADDING FORCES VECTORIALLY

In this section we consider what happens when more than one force acts on an object at the same time. Earlier, we stated that forces behave mathematically like vectors; we now want to verify this claim by investigating a situation where multiple forces act in two dimensions.

Before getting started, recall that a two-dimensional vector $\vec{F}$ can be written as

$$\vec{F} = F_x\hat{x} + F_y\hat{y}$$

where $F_x$ and $F_y$ are the $x$ and $y$ components of the vector, while $\hat{x}$ and $\hat{y}$ are the unit vectors pointing in the $x$ and $y$ directions. Of course, this vector can also be written in terms of its magnitude and direction, where the magnitude is given by $F = |\vec{F}| = \sqrt{F_x{}^2 + F_y{}^2}$, and the direction is an angle specified with respect to the given coordinate system.

In the next activity, we consider three forces acting on a single object, for which you will need:

- 1 data-acquisition system
- 2 electronic force sensors
- 2 table clamps

- 2 90° rod clamps
- 2 long rods and 2 short rods
- 1 hanging mass (approximately 500 g)
- 1 circular book ring (about 2 inches in diameter)
- 3 small pieces of string or thread
- 1 protractor

### 5.6.1. Activity: Verifying That Forces Behave Like Vectors

**a.** Set up the software so you can measure two force sensors simultaneously and check that they are both set to the 10-N scale. After zeroing the force sensors with the hooks pointing down, begin taking data and hang the mass from each sensor separately for a few seconds. These readings tell us the amount of force the mass is exerting on the sensor. Record the (average) force readings below for each sensor. If the sensors don't measure approximately the same value (within a few tenths of a newton) ask your instructor for help!

**b.** Using the clamps and rods, set up the situation shown in Fig. 5.6. You want the mass to hang straight down, one of the force sensors to pull horizontally, and the second force sensor mounted so that it pulls at an angle $\theta$ between 45° and 60° with respect to the horizontal. Once you have everything set up correctly, place your hand under the hanging mass and lift it very slightly so that there are (essentially) no forces being exerted on the force sensors. While holding the mass up, zero the force sensors. Then start the experiment and gently release the mass so that it hangs freely. Report the (average) force values measured by the sensors below, along with the angle $\theta$ measured using a protractor.

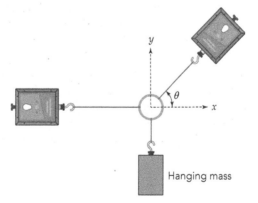

**Fig. 5.6.** An experiment to verify the vector nature of forces.

**c.** Make a sketch that shows the three force vectors acting on the ring in their correct orientation (your sketch should look quite similar to Fig. 5.6). Draw the force vectors using a scale of 0.5 cm per newton and label each force with its magnitude. Then, using the coordinate system shown in Fig. 5.6, write each vector in component form (don't forget units and unit vectors). **Note**: Use your result from part (a) for the force from the hanging mass.

**d.** The *total*, or *net*, force acting on the ring is simply the sum of these three vectors. We can do this *vectorially*, by drawing the three vectors head-to-tail, or *algebraically*, by summing the $x$ and $y$ components. Determine the net force using both of these methods and show that you get the same answer—approximately zero—either way.[3]

**e.** Recall that when there is a single force acting on an object, we observed that the object's acceleration is proportional to this force. However, in this experiment there are multiple forces acting on the ring, and the ring is not accelerating at all. How do you explain this observation?

---

In the previous activity, multiple forces are acting on a single object (the ring) and yet the object remains at rest with *zero* acceleration. While this might seem odd initially, you should have found that the *net* force acting on the object is zero (within our measurement uncertainty). The fact that forces behave like vectors means there is no mathematical difference between a situation in which several forces act on an object simultaneously and a situation in which a single force (equal to the net force) acts on that same object. Therefore, we conclude that **the acceleration caused by *several* forces acting on an object is proportional to the *net* force acting on that object**. In other words, the acceleration is determined by the *net* force:

$$\vec{a} = \frac{\vec{F}^{\text{net}}}{m} \tag{5.1a}$$

This expression can also be written in component form as

$$a_x = \frac{F_x^{\text{net}}}{m} \tag{5.1b}$$

---

[3] Note that we are assuming the gravitational force acting on the ring is negligible (much smaller in magnitude compared to the other forces) in this situation. This will be a good approximation as long as the mass of the ring is much smaller than the hanging mass.

with similar expressions for $a_y$ and $a_z$. Although this may not seem like a major step forward, it turns out to be a very powerful idea that we will use again and again throughout this course.

It's worth emphasizing that the net force can be found by adding forces vectorially (or by adding their components algebraically). Mathematically, we write

$$\vec{F}^{\text{net}} = \sum \vec{F} = \vec{F}_A + \vec{F}_B + \cdots \qquad (5.2a)$$

to denote vector addition, where $\vec{F}_A, \vec{F}_B$, etc. are the individual force vectors. In words, the *net* (or total) force vector is equal to the sum of the individual force vectors. Notice that there are no minus signs here; we are *adding* the vectors. Vectors can be thought of as arrows, and we add them *vectorially* by placing the arrows head-to-tail.

Alternatively, we can add the components *algebraically*, which we write as

$$F_x^{\text{net}} = \sum F_x = F_{Ax} + F_{Bx} + \cdots \qquad (5.2b)$$

Here, $F_x^{\text{net}}$ represents the *net* force in the $x$-direction, which is the algebraic (regular) sum of the $x$ components of all the forces that are acting. There are no minus signs here either; we simply *add* the components of all the forces that are acting. Of course, the vector components themselves can be positive or negative (with signs determined by the directions of the forces relative to the coordinate system), so once these components are substituted into Eq. (5.2b) there may be some minus signs present.

In practice, we almost always sum vector components, as in Eq. (5.2b), rather than try to add the actual vectors, as in Eq. (5.2a). The two procedures are equivalent, but it's a lot easier to add numbers than it is to add arrows (even if some of those numbers are negative). When working on problems that require two or three dimensions, we must consider the $y$ and $z$ directions as well:

$$F_y^{\text{net}} = \sum F_y = F_{Ay} + F_{By} + \cdots \qquad (5.2c)$$

$$F_z^{\text{net}} = \sum F_z = F_{Az} + F_{Bz} + \cdots \qquad (5.2d)$$

Finally, it is worth mentioning that some textbooks refer to the *net* force as a *combined* or *total* force, while other textbooks may talk about the *resultant* force. All of these different terms refer to the same thing.

### 5.6.2. Activity: Calculating the Net Force

**a.** Imagine an object that is being acted on by two forces. Force $\vec{F}_1$ has a magnitude of 5.0 N and acts up and to the right, making an angle of 45° with the positive $x$-axis. Force $\vec{F}_2$ acts down and to the left, with a magnitude of 2.5 N and making an angle of 30° below the negative $x$-axis. Draw a diagram of this situation showing a small box with forces $\vec{F}_1$ and $\vec{F}_2$ acting in the specified directions (using 1 cm of length per newton). Sketch a standard coordinate system to the right of your diagram.

**b.** Calculate the net force acting in both the $x$ and $y$ directions using vector *components*. Also, show the vector addition with arrows. Be sure to show your work and include proper units.

**c.** Briefly describe the motion of this object qualitatively. In particular, will the object accelerate, and if so, in what direction?

## 5.7 WHAT HAPPENS WHEN THE NET FORCE IS ZERO?

In this section we consider an important special case in more detail—the situation in which the net force acting on an object is equal to zero.

### 5.7.1. Activity: Facts About Zero Net Force and Motion

**a.** Consider a low-friction cart on a track. Suppose we apply a force $\vec{F}_A$ to the cart that acts to the right and another force $\vec{F}_B$ of *equal* magnitude that acts to the left. Draw a diagram of the cart showing the two forces and choose (sketch) a coordinate system. Determine the net force $F_x^{net}$ acting in the $x$-direction.

**b.** Given this net force, what is the acceleration $a_x$ of the cart (in the $x$-direction).

c. Now consider a completely different cart on a low-friction track. What is the acceleration of the cart if the cart is:

1. At rest?

2. Moving with a constant speed in the positive direction?

3. Moving with a constant speed in the negative direction?

d. For each of your answers in part (c), what is the net force acting on the cart? Explain briefly.

---

When the net force acting on an object is zero, the object's acceleration will be zero. Conversely, if an object's acceleration is zero, then the net force acting on it must be zero. These statements are a direct result of the relationship between force and motion shown in Eq. (5.1); namely, that the acceleration of an object is proportional to the net force acting on that object. But notice that while an object with zero net force acting on it has an *acceleration* that's zero, Eq. (5.1) says nothing about the velocity (or position) of the object. In particular, an object moving with a (non-zero) constant velocity will have zero acceleration, which means the net force acting on such an object must be zero.

We would like to experimentally investigate the motion of a low-friction cart that has zero net force acting on it. To undertake this investigation, we will use two equal hanging masses to apply equal magnitude forces in opposite directions[4] (see Fig. 5.7).

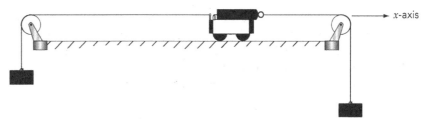

**Fig. 5.7.** Setup showing a low-friction cart rolling on a level track as two hanging weights combine to exert zero net force on the cart.

This activity can be done as a class demonstration, and requires the following materials:

- 1 high table
- 1 smooth ramp or level surface 2 m long
- 1 low-friction dynamics cart
- 2 lengths of string, 2 m
- 2 low-friction pulleys
- 2 hanging masses, 500 g or 1 kg

---

[4] Although we won't quantify forces due to such hanging weights until Unit 6, you can assume that equal hanging masses on each side will result in equal and opposite forces on the cart.

Before making any observations, be sure to level the track and balance the masses.

---

### 5.7.2. Activity: Motion with Zero Net Force

**a.** Suppose the cart in Fig. 5.7 is being held at rest when the two hanging masses are at exactly same height. What do you think will happen when the cart is released? What if the cart is held at rest when the right mass is at a higher location than the left mass and then released? Record your predictions below.

**b.** Now try the experiment and record your observations. What can you conclude about the heights of the masses in this experiment?

**c.** Now suppose the cart in Fig. 5.7 receives a quick push that starts it moving in one direction. What do you think its motion will be like *after* the push (assuming there is no friction acting on the cart)? Will it speed up, slow down, or move with a constant speed?

**d.** Describe what you observe when this experiment is carried out. Is the velocity of the cart (approximately) constant? If you notice any acceleration, is it consistent with there being a small amount of friction in the system? What do you think the motion would be like if friction could be completely eliminated? Explain briefly.

**e.** The observation you just made should enable you to state *Newton's First Law of Motion*. Finish the following statement in a manner that's compatible with your observations.

**Newton's First Law:** If an object moving at a constant velocity experiences no net force (including friction), it will …

**Remarks About Newton's First Law**

Newton's first law is significant because it allows an observer who is moving at a constant velocity with respect to another observer to discover the same laws of motion. This is important because we are, after all, living on a planet that is in constant motion. To see how this might come into play, suppose you observe that a cart is at rest in the laboratory while your partner measures the position of the cart while walking away from you at a constant velocity. Clearly, your partner will observe the cart moving away with a constant velocity. Although you and your partner will observe different velocities, you both agree that the cart is not *accelerating* and is therefore experiencing *no net force*! Non-accelerating reference frames—those that are moving with a constant velocity—are called *inertial* reference frames, and scientists have found that the laws of physics are the same in all inertial reference frames.

## 5.8  DEFINING AND MEASURING MASS

**Fig. 5.8.** Would you rather push a small sports car or a large van?

Imagine a good friend calls you in a panic. Their car battery is dead, and they need you to come outside and help push their car to get it started. You must get the car moving at a speed of about 12 mph so they can "pop the clutch" and start the engine.[5] You casually answer sure, you'll be right out. Then you remember that your friend owns two vehicles—a large delivery van and a smaller sports car (Fig. 5.8). Which vehicle do you hope your friend wants to start?

Although you won't need a small car and large van, it would be helpful to have the following items available for this activity:

- 1 lightweight ball (e.g., a soccer ball)
- 1 bowling ball (or other heavy object)

### 5.8.1.  Activity: Causing a Car to Accelerate

**a.**  Which vehicle do you think would be easier to push from rest to a speed of about 12 mph, a small car or a large van? Explain your answer (e.g., what does it mean to be "easier" to push?).

**b.**  The physical property that's at play here is called *inertia*. Inertia is a *resistance* to changes in motion—a resistance to acceleration. Imagine you are going to kick a ball across a field to a friend. Would you rather kick a soccer ball or a bowling ball? Why?

---

[5] This technique of starting a car is known as *compression starting* and can only be done with manual transmission vehicles powered by an internal combustion engine. If your car battery dies unexpectedly, compression starting is a great way to get the car started so you can drive to a service station to get the battery replaced. As of the writing of this text, these types of vehicles are becoming less and less common, so it is likely only a matter of time before compression starting becomes a thing of the past.

**c.** To get a physical feel for the concept of inertia, pick up a soccer ball (or some similar type of object) and move it side-to-side (left and right) as fast as you can. Now pick up a bowling ball and do the same thing. Explain your observations below.

**d.** What characteristic of an object seems to determine how much inertia it has? Explain briefly.

## Inertial Mass

Most people think an object's inertia is related to how much "stuff" is in the object. Of course, exactly what one means by "stuff" is not clear. One way to define the amount of "stuff" is to focus on the property of *inertia*, an object's resistance to acceleration. Notice that the constant of proportionality (the slope) in the force versus acceleration experiment has exactly this property. In other words, a large slope signifies that a large force is needed to accelerate the object.

Therefore, we can define an object's *inertial mass*, denoted by the symbol $m_I$, to be the constant of proportionality between $F_x^{net}$ and $a_x$. Mathematically, $m_I \equiv F_x^{net}/a_x$, where $F_x^{net}$ is the net force applied to the object, and $a_x$ is the object's observed acceleration (all in the $x$-direction). The inertial mass literally tells us how many newtons of force are needed to accelerate the object at a rate of $1\,m/s^2$, and our (one-dimensional) law of motion becomes $F_x^{net} = m_I a_x$. This definition of (inertial) mass is focused on the property of inertia (hence the name).

## Another Definition of Mass

The inertial mass may seem like a strange way of defining how much "stuff" is in an object, but it is a perfectly reasonable way of defining mass. Interestingly, and perhaps surprisingly, there are other ways of defining mass. In fact, you probably already have a different way of thinking about the mass of an object.

### 5.8.2. Activity: Ideas About Mass and Its Measurement

**a.** Can you think of another way of defining *mass* that is different from how we defined the inertial mass? **Hint**: How could you determine whether two different objects have the same mass?

**b.** Suppose you find that two objects have the same mass according to your definition above. How could you determine whether a third object has twice the mass of either of the original objects?

### The Gravitational Mass

Another way to define mass relates how the objects are affected by gravity (technically, the *gravitational force*). One time-honored method people have used to compare the masses of two objects is to put them on a "balance" (Fig. 5.9). If the two objects balance, we say that they have the same mass.

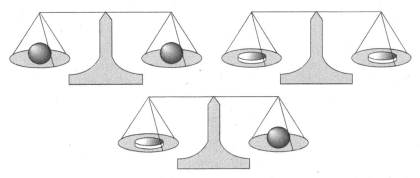

**Fig. 5.9.** A common method of determining mass that assumes two objects have the same mass if they experience the same gravitational force.

As you may be aware, what we are really doing here is comparing the gravitational forces acting on the two objects. If the gravitational forces are the same on the two objects, we assert that the two objects must have the same mass (the same amount of "stuff"). Notice that this definition of mass relies on the strength of the gravitational force instead of the relation between force and acceleration, as with our earlier definition. Thus, to distinguish this quantity from our earlier definition, we might call it the *gravitational mass* and give it the symbol $m_G$.

Defining mass in this way requires that we develop a *standard* mass. Once this standard has been developed, then all other masses can, at least in principle, be compared to the standard mass using a balance. An object has the same mass as the standard mass if it balances the standard mass on a gravitational scale, while an object has twice the mass of the standard mass if it takes two standard masses to balance the object on a gravitational scale, etc. In this way, the gravitational mass of any object can be determined.[6]

In the next section, we will consider in more detail these two ways of defining mass, including whether or not they are equivalent.

## 5.9   COMPARING OUR DEFINITIONS OF MASS

We have already verified a law of proportionality between a force and the acceleration of an object when little friction is present. We then extended this law to account for multiple forces acting on the object. For motion along the $x$-axis, this law can be expressed in the form

$$F_x^{\text{net}} = ma_x \tag{5.3}$$

where $F_x^{\text{net}}$ is the *net* force exerted on an object, $a_x$ is the acceleration, and $m$ is the constant of proportionality (the slope of the graph of $F_x^{\text{net}}$ versus $a_x$).

We saw that this constant of proportionality represents a *resistance* of an object to acceleration, something we referred to as the inertial mass $m_I$. Activity 5.8.1 provided a concrete example of inertia; we saw that it takes a lot more force to accelerate a bowling ball compared to a soccer ball. Thus, we would say that the bowling ball has more *inertial* mass than the soccer ball. Notice that this resistance to acceleration has nothing to do with gravity. That being said, the bowling ball is a lot heavier than the soccer ball, so we would also conclude that the bowling ball has more *gravitational* mass than the soccer ball. Thus, even though the inertial mass and gravitational mass are defined quite differently, at first glance it seems as though they might be related.

To explore this idea further, we will investigate the forces and accelerations that arise when we push and pull rolling carts having different gravitational masses (this investigation is basically an extension of the one we undertook in Section 5.5). For the activities in this section, you will need:

- 1 electronic scale
- 1 low-friction dynamics cart
- 2 mass bars, 500 g (to add mass to the cart)
- 1 cart track or level surface, 2 m long
- 1 data-acquisition system
- 1 ultrasonic motion sensor
- 1 force sensor

In preparation for this activity, attach a force sensor with a hook on its end firmly to the cart. Make sure the force sensor is set to the 10-N scale and place the cart on a level track or surface (see Fig. 5.10). At this point you should *not* include any of the mass bars for the measurement.

---

[6] A gravitational balance is not the only way to measure gravitational mass. If we hang an object from a spring scale, the spring will stretch due to the gravitational force acting on the object. We can then compare this force to the force measured when the standard mass is hung from the same spring scale (one can do the same thing using an electronic scale or force sensor). Because spring scales and electronic scales are so convenient to use, they have largely replaced the old-fashioned gravitational balance. They still, however, rely on the gravitational force when comparing the masses of objects, so they still measure gravitational mass.

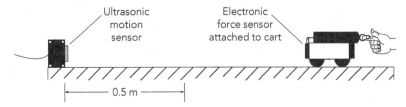

**Fig. 5.10.** Setup showing a motion sensor tracking the acceleration of a cart rolling on a level track as a force sensor detects the pushes and pulls on it. As usual, the cart should remain roughly 0.5 m away from the motion detector at all times.

Set up the motion software to display three graphs:

1. Force versus time
2. Acceleration versus time
3. Force versus acceleration

### 5.9.1. Activity: Measuring Force versus Acceleration

**a.** As in Section 5.5, begin by *zeroing the force sensor* and then record data as you firmly grasp the force sensor hook and smoothly push and pull the cart back and forth on the track. Sketch the force and acceleration graphs you observed. **Note**: We want the force and acceleration graphs to be pretty smooth, so you may need to try the experiment a few times to get clean data (the results tend to be better if you push and pull with a maximum force of about 5 N). In addition, make sure the force-acceleration graph is a *scatter plot* such that the points are *not* connected.

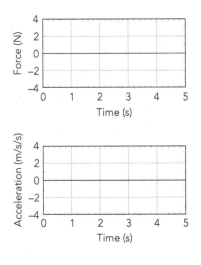

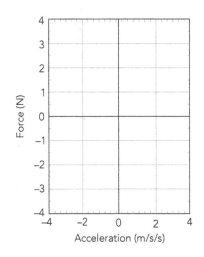

**b.** As before, the data for the force-acceleration graph should look like a straight line going through the origin. Using the motion software, fit a line to the data and determine its slope. Write this value below, remembering to include units. This slope is what we defined as the *inertial mass $m_I$.*

**c.** To determine the gravitational mass $m_G$, place the cart (with the force sensor attached) on an electronic scale. When weighing the cart, be careful with the cable; since most of the cable was not being accelerated, the cable should not be included when you weigh the cart (just put it on the table next to the scale). Report your value for $m_G$ below, remembering to include units.

**d.** How does the gravitational mass $m_G$ compare to the inertial mass $m_I$? Do these values seem to be related?

---

### Increasing the Gravitational Mass

Now we're going to repeat the above experiment after placing additional mass on the cart. Specifically, we will add two 500-g (gravitational) mass bars to the cart.

#### 5.9.2. Activity: Increasing the Gravitational Mass

**a.** If you repeat the above experiment with an additional 1 kg of (gravitational) mass, how do you think the force-acceleration graph will change? Will it still be a straight line? Will the slope change? If so, will it increase or decrease? Be as specific as possible.

**b.** As before, *zero the force sensor* and then record data as you *smoothly* move the cart back and forth on the track. Sketch the force and acceleration graphs you observe.

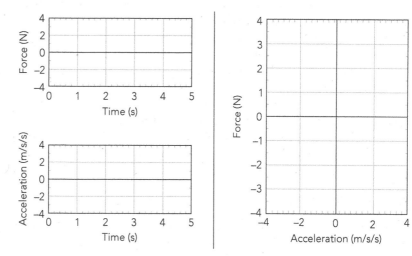

**c.** Using the linear fit feature in the motion software, indicate the value of the slope of the force-acceleration graph (include units). This is a measurement of the inertial mass $m_I$.

**d.** Now determine the gravitational mass of the cart, force sensor, and mass bars and report this value below. **Note**: It is probably easier to simply add the (gravitational) mass of the bars to the value you found in the last activity for the cart and force sensor.

**e.** How does the gravitational mass compare to the inertial mass this time? In light of the two experiments you just performed, what is the apparent relationship between the inertial mass and the gravitational mass? Explain briefly.

**The Equivalence Principle**

Hopefully, the last activity demonstrated that, within the limits of experimental uncertainty, *the inertial and gravitational masses are the same*. Even if you are not impressed, you should at least be aware that it is *not* obvious that these two definitions of mass—gravitational and inertial—should lead to the same value. In fact, this equivalence is *assumed* in Newton's theory of gravity and is known as the *equivalence principle*. Extremely precise experiments have been performed demonstrating that there is no measurable difference between the two types of mass to less than one part in $10^{14}$.

Albert Einstein believed that this equivalence was not a coincidence, and it led him to develop a different theory of gravity known as *general relativity*. In this theory, Einstein suggests that gravity is not a force at all, but instead a curvature of what he called *spacetime*. Although general relativity is now considered a superior theory of gravity compared to Newton's law of gravitation, the two theories agree almost perfectly for any situation you are likely to experience in the everyday world. It is only near extremely massive objects that we begin to see differences between the two theories.

### Newton's Second Law

Now that we have found that the inertial mass and gravitational mass have the same value, we can stop distinguishing between them and simply define "the mass" as either the inertial or gravitational mass: $m \equiv m_I = m_G$. In practice, it is far easier to find the mass of an object using some type of gravitational scale. It is useful to keep in mind, however, that mass has both inertial and gravitational properties; that is, mass resists acceleration while also being attracted to Earth.[7]

For completeness, we now rewrite Newton's second law (in one-dimension) as

$$F_x^{\text{net}} = \sum F_x = ma_x \qquad (5.4)$$

where $F_x^{\text{net}}$ is the net force (in the $x$-direction) on the object, $a_x$ is the acceleration (in the $x$-direction), and $m$ now represents the (gravitational or inertial) *mass* of the object.

## 5.10   SI UNITS AND SUMMARIZING NEWTON'S FIRST AND SECOND LAWS

As you may already know, the Systemè Internationale, or SI, system of units provides us with a standard set of units. This system was formally established in 1960 to provide a standard system of units that all scientists throughout the world can use. Prior to Unit 5, we really only required two quantities: length and time. Now that we have defined mass, we are ready to formalize the three primary units used in mechanics. The SI units for these fundamental quantities are shown below with their pre-2019 definitions[8] (see Fig. 5.11).

**Fig. 5.11.** Three of the primary SI units that are most common when studying force and motion are the meter (length), the second (time), and the kilogram (mass). The other units we will encounter are derived from these basic quantities.

---

[7] In addition to being attracted to Earth, a mass is gravitationally attracted to all other masses. However, the gravitational attraction is extremely weak and is only noticeable when near an object with an extremely large mass, such as Earth.

[8] The SI system underwent a significant change in 2019 that resulted in five constants of nature being re-defined as having exact values: the speed of light, the Planck constant, the elementary charge, the Boltzmann constant, and the Avogadro constant. These changes resulted in a number of quantities being re-defined (the kilogram, ampere, kelvin, and mole). The new definitions are made so that each unit "is unique and provides a sound theoretical basis on which the most accurate and reproducible measurements can be made." Unfortunately, the modern definitions are sometimes difficult to fully comprehend, so we stick with the pre-2019 definitions even though they are no longer technically true.

**Fig. 5.12.** The (pre-2019) SI force unit expressed in terms of length, mass, and time: A newton (N) is the SI unit of force, defined as the force that, when acting on a 1-kg mass, causes an acceleration of 1 m/s/s.

### SI Units for Mechanics (Pre-2019 Definitions)

*Length:* A **meter** (m) is the distance traveled by light in a vacuum during a time of 1/299,792,458 second.

*Time:* A **second** (s) is defined as the time required for a particular light wave given off by a cesium-133 atom to undergo 9,192,631,770 oscillations.

*Mass:* A **kilogram** (kg) is defined as the mass of a platinum-iridium alloy cylinder kept in a special chamber at the International Bureau of Weights and Measures in Sévres, France.

The electronic balance and spring scales often used in laboratories have been calibrated using replicas of the "true" standard kilogram mass kept in a vault in France. These fundamental units and Newton's second law can also be used to define the newton as a unit of force. One newton of force is defined as the force that, when acting on a 1 kg mass, causes an acceleration of 1 m/s$^2$ (see Fig. 5.12). This means that the newton can be written in terms of the primary SI units as

$$1\,\text{N} = 1\,\frac{\text{kg}\cdot\text{m}}{\text{s}^2}$$

### Summarizing Newton's Laws

The main purpose of Unit 5 has been to explore the relationships between forces on an object, its mass, and its acceleration. Along the way we developed Newton's first two laws of motion, which we now ask you to summarize in your own words.

---

#### 5.10.1. Activity: Newton's Laws in Your Own Words

Express Newton's first and second laws of motion in your own words.

- The First Law:

- The Second Law:

---

As you may have noticed, Newton's first law is really just a special case of Newton's second law. Thus, the second law is usually considered the more important of the two; if you remember the second law, the first law follows by setting $F_x^{\text{net}} = 0$.

### Newton's Second Law as a Vector Equation

All of the experiments we performed to arrive at Newton's second law were in one-dimension. However, because each component of a vector is independent of all other components, we can immediately generalize this law to a vector equation:

$$\vec{F}^{\text{net}} = \sum \vec{F} = m\vec{a} \tag{5.5}$$

Written in this manner, the equation is valid independent of any coordinate system, which means that we are free to choose any coordinates we desire. This flexibility can be used to choose a coordinate system that simplifies the problem as much as possible. For example, if the situation turns out to be one-dimensional, we might choose the $x$-axis to lie along the direction of motion, thereby turning the situation from an inherently three-dimensional problem into a one-dimensional problem.

Once a coordinate system has been chosen, the vector form of Newton's second law is equivalent to three separate component equations that are valid along the three coordinate axes:

$$F_x^{\text{net}} = \sum F_x = ma_x$$

$$F_y^{\text{net}} = \sum F_y = ma_y$$

$$F_z^{\text{net}} = \sum F_z = ma_z$$

**Note**: Students sometimes get confused by Newton's second law and end up thinking of the quantity on the right (mass times acceleration) as if it were a force. To help avoid this type of confusion, we suggest that the actual forces are always placed on the left side of the equation. Once all the forces have been accounted for (and properly added), you then set this net force *equal* to $m\vec{a}$, which is written on the right side of the equation. We will have much more to say about this in the units to come.

### Final Comments on Force, Mass, and Motion

We started our study of Newtonian dynamics by developing the concept of force. When asked to define force, most people initially think of a push or pull, such as the tug of a rubber band. By studying the acceleration that results from a force when little friction is present, we found a simple relationship between force and acceleration. However, pushing on a wall doesn't seem to cause the wall to accelerate. Similarly, an object dropped close to the surface of Earth accelerates, and yet there is no visible push or pull on it.

Newton recognized that he could define the *net* force as the force that causes acceleration. He reasoned that if the obvious applied forces did not account for the observed acceleration, then other "invisible" forces must be present (Fig. 5.13). A prime example of an invisible force is that of "gravity," the attractive force pulling objects toward Earth. Identifying the forces acting on an object can sometimes be challenging, particularly when some of them are *passive* forces. Passive forces only act in response to either the motion of an object or other forces acting on the object. For example, pushing on a wall results in a passive force. Assuming the wall stays put (its acceleration remains zero), the net force on the wall must be zero. Therefore, if you push on the wall in one direction, there must be some other force acting on the wall such that the net force on the wall is zero. Furthermore, this passive force acting on the wall will "disappear" the instant you stop pushing so that the net force on the wall remains zero.

**Fig. 5.13.** Newton recognized that the laws of motion discovered through the use of applied forces could also be used to discover the nature of gravitational forces and the forces of friction.

Friction forces are another example of passive forces. The passive nature of friction becomes clear when you think of the person sliding along the floor at constant velocity under the influence of an applied force (as we saw in the beginning of this unit). According the Newton's First Law, an object moving at a constant velocity must have no net force acting on it. Newton reasoned that the applied force in one direction had to be opposed by a friction force acting in the opposite direction to "oppose the motion." But what happens if the applied force is removed? The friction force causes the sliding person to slow down and stop, but then it disappears. If the friction force continued to act, the person would eventually start speeding up in the opposite direction (which clearly doesn't happen!).

During the rest of our study of the Newtonian formulation of classical mechanics, your task will be to work with different types of forces to predict and explain observed motions using Newton's laws. In the next unit we will use Newton's laws to understand what happens to an object when dropped close to the surface of Earth. We will also learn to employ Newton's laws in two-dimensional situations when the motion of an object and the forces that act on it do not all lie along the same line.

:· **5.11** **PROBLEM SOLVING**

### 5.11.1. Activity: Car Acceleration

One of the advantages of an electric vehicle (EV) over one using an internal combustion engine (ICE) is that an electric motor does not require a traditional transmission, nor does it suffer from a lag as the engine rpm ramps up from idle. The result is that an EV provides nearly instant power to the wheels and is therefore capable of producing very high accelerations compared to traditional ICE vehicles. For example, a Tesla Model S P100D with so-called "ludicrous mode" has an acceleration of $10.7 \, \mathrm{m/s^2}$, a mark that is surpassed by only a handful of exotic (and extremely expensive) ICE vehicles. For the following problem we will assume this acceleration is constant.

Hadrian/Shutterstock.com

a. Assuming the Tesla has a mass of 2250 kg, determine the *net force* needed to accelerate the Model S. (In reality, friction and air resistance will act in the opposite direction of motion, so the force produced by the car will need to be even larger to achieve this acceleration.)

b. Determine how much time (in seconds) it takes the Model S to go from zero to 60 mph.

c. Now, assume the Tesla is occupied by four adults, each of whom weighs 180 pounds. They are also carrying four suitcases, each of which weighs 40 pounds. If the car produces the same force as it did before, what is the new acceleration? How much longer (in seconds) will it take to reach 60 mph?

### 5.11.2. Activity: Rocket Acceleration

A Saturn V rocket fully loaded with fuel has a mass of $2.90 \times 10^6$ kg and a weight (on Earth) of $6.39 \times 10^6$ pounds (1 N = 0.22 pounds). Note that over 90% of the rocket's initial mass consists of fuel. The upward acceleration of the rocket during the early portion of flight is $2.30\,\text{m/s}^2$.

NASA

a. Determine the upward force supplied by the rocket engine (the engine's *thrust*) during the early portion of flight. Use SI units (meters, seconds, kilograms, and newtons) for all your calculations and ignore the effect of air resistance. **Hint**: Begin by drawing a diagram for the situation showing the forces, recognizing there are only two forces acting on the rocket.

b. Roughly 2 minutes into the flight, the acceleration has increased significantly, approaching the maximum amount a human body can handle. Yet it is the same engine supplying the same amount of thrust. How is this possible? What about the rocket has changed since it first took off? (There are actually a few effects that contribute, but one of them is most important.)

c. Assume that at this point in the flight the rocket has a mass of $8.99 \times 10^5$ kg and a weight (on Earth) of $1.98 \times 10^6$ pounds. What is the acceleration of the rocket now?

# UNIT 6: GRAVITY AND PROJECTILE MOTION

*This beautiful image captures an explosion at Tungurahua volcano in Ecuador on May 31, 2010. The parabolic trajectories followed by the glowing projectiles are a characteristic shared by all objects moving under the influence of a constant force. In this unit, we investigate the gravitational force near the surface of Earth and explore the motion of objects under the influence of this force. Photo credit: Benjamin Bernard/Wikimedia Commons/CC BY 4.0, via Wikimedia Commons.*

# UNIT 6: GRAVITY AND PROJECTILE MOTION

## OBJECTIVES

1. To explore the gravitational force and study the nature of motion along a vertical line near Earth's surface.

2. To use Newton's laws to uncover "invisible" forces for describing phenomena such as gravity and connect the gravitational force to the concept of weight.

3. To understand the experimental and theoretical basis for describing projectile motion as the superposition of two independent motions: (1) an object falling in the vertical direction, and (2) an object moving in the horizontal direction with no net force.

4. To investigate the gravitational force far from Earth's surface.

## 6.1 OVERVIEW

When an object is dropped near the surface of Earth, there's a force that pulls it to the ground. However, it's not obvious exactly what's responsible for this force. After all, there is nothing touching the object (other than air, and it doesn't seem likely that the air is pulling the object to the ground). Most people casually refer to the cause of this falling motion as the action of "gravity." But what exactly is gravity? Can we describe its effects mathematically? Can Newton's laws be used to predict the motion caused by gravity?

In the first few activities of this unit, we will study the phenomenon of gravity near the surface of Earth by examining an object's vertical motion. In addition, we will connect the *gravitational force* to the concept of weight. We will then study *projectile motion*, in which an object is launched at some angle with respect to the horizontal and undergoes two-dimensional motion. Finally, we will briefly examine the law of universal gravitation, which describes the gravitational force even when one is *not* close to the surface of Earth.

## VERTICAL MOTION

### 6.2 DESCRIBING HOW OBJECTS FALL

Let's begin our study of gravity by observing the motion of two falling objects (Fig. 6.1). For these activities, you will need:

- 1 small rubber ball
- 1 flat-bottomed coffee filter

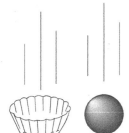

**Fig. 6.1.** How do a coffee filter and a ball fall when they are dropped?

#### 6.2.1. Activity: Predicting Falling Motions

**a.** *Predict* how the ball will fall when dropped near the surface of Earth, providing as much detail as possible. For example, will it fall with a constant velocity? Will it speed up quickly to some final velocity and maintain this velocity as it continues to fall? Will it speed up the entire time? Will it undergo some other type of motion? Try to provide a reason for your prediction.

**b.** Similarly, *predict* how the coffee filter will fall when dropped near the surface of Earth, providing as much detail as possible. If the two objects are dropped at the same time, which one do you think will hit the ground first? Explain the reason for your prediction.

**c.** Finally, *predict* how the coffee filter will fall if it is crumpled up into a little ball. Will its motion differ from how it falls when uncrumpled? If so, how will its motion compare to that of the rubber ball? Explain briefly.

**d.** Now perform the experiments and observe what happens. Drop the rubber ball and the (uncrumpled) coffee filter at the same time. Then repeat the experiment using the crumpled coffee filter. Describe your observations and compare to your predictions. If your observations differed from your predictions, try to explain why.

When the uncrumpled coffee filter and ball are dropped at the same time, the fact that the ball hits the ground first probably came as no surprise. In fact, you may have been able to see that the coffee filter falls at a reasonably constant velocity. On the other hand, you may have had trouble observing the ball's motion in real time because it happens very quickly.

As you probably guessed, the uncrumpled coffee filter is significantly affected by *air resistance*. However, when the coffee filter is crumpled up, it reaches the ground in roughly the same amount of time as the ball. That's because air resistance is much smaller for the crumpled coffee filter. It is easier to describe the motion of an object when air resistance is negligible, so this is where we will focus our attention.

## 6.3   DESCRIBING HOW OBJECTS RISE AND FALL

Let's continue our study of how the gravitational force (near Earth's surface) affects the motion of an object by returning to an experiment we did in Activity 4.3.3: the motion of an object that is tossed up vertically. As you may remember, we looked at the acceleration of a ball during three different portions of the motion: rising, at the highest point, and falling. For the following activity, you can use the same data from Activity 4.3.3. Alternatively, you can use a motion sensor to record position-time data for a tossed ball, or you can analyze a video of a tossed ball. Here, we describe the motion sensor approach. The following equipment may be helpful:

- 1 basketball (or similar object)
- 1 data-acquisition system
- 1 ultrasonic motion sensor
  OR
- video of tossed ball

Fig. 6.2. A ball being tossed into the air.

### 6.3.1.  Activity: Predicting the Motion of a Tossed Ball

a. Toss a ball straight up in the air a couple of times and observe its motion (Fig. 6.2). Assuming the upward direction is positive, predict in the table below whether the velocity and acceleration are positive, negative, or zero during each part of the motion. **Note**: Don't go back and look at your previous answers from Unit 4; just answer the question as best you can using your current understanding of the situation.

|              | Moving up | At the highest point | Moving down |
|--------------|-----------|----------------------|-------------|
| Velocity     |           |                      |             |
| Acceleration |           |                      |             |

b. Do you expect the acceleration when the ball is rising to be different in some way than the acceleration when the ball is falling? Why or why not?

**c.** What do you think the acceleration will be when the ball is at its highest point? Why?

---

### Using a Motion Sensor to Track the Ball's Motion

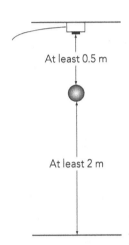

At least 0.5 m

At least 2 m

1. Use some rods and clamps to set up the motion sensor so that it is as high as possible and facing directly downward. If the sensor has both a "cart" and "ball" mode, set it to ball mode. In addition, change the direction of the motion sensor axis so that upward is positive and then zero the sensor so that the ground is considered the origin.
2. Set the motion software to record 30 points/second.
3. Hold the ball at least 50 cm below the motion sensor and start collecting data, dropping the ball straight down and allowing it to bounce on the ground.
4. Check to see if the motion sensor successfully tracked the ball throughout its motion. We are interested in the motion of the object *after the first bounce* and *before the second bounce*. The position graph for this portion of the motion should be a smooth curve that shows the ball moving up and then coming back down. Repeat the experiment as needed to get clean results.

**Note**: Although we are still only considering one-dimensional motion in this section, it is customary to reserve the $x$-axis for horizontal motion and the $y$-axis for vertical motion. Thus, vertical position, velocity, acceleration, and force components are typically denoted as $y$, $v_y$, $a_y$, and $F_y$, respectively. However, the motion sensor data will likely show the position as $x$ because the program has no idea how the motion sensor is oriented.

---

#### 6.3.2. Activity: Analyzing the Motion of a Tossed Ball

**a.** Look at a graph of the vertical position as a function of time ($y$ vs $t$). Adjust the scale on the time axis to show only the portion of the motion where the ball is in the air after the first bounce (or after the toss if you threw the ball into the air). Affix a copy or sketch the graph in the space below. Be sure to label your axes.

**b.** Based *only* on the position-time graph, can you tell whether or not the ball has a *constant* acceleration throughout its motion? Explain briefly.

**c.** Based on our previous experiments, it is perfectly reasonable to guess that this object has a constant acceleration; unfortunately, it's nearly impossible to tell just by looking at the position-time graph. One way to determine if the acceleration is indeed constant is to fit the data using an appropriate equation describing the (vertical) position as a function of time. What type of function should we use to model the position-time graph if the acceleration is constant? Briefly explain why.

**d.** Now perform a fit to the data. **Note**: Because the region of interest doesn't start at $t = 0$, you should allow for a time offset in your fit. You should find that a quadratic function leads to a pretty good fit, which supports the hypothesis that the acceleration is constant. Write down the fit equation below, and from this equation determine the initial position, initial velocity, and the (constant) acceleration. Record these values below, being sure to include proper signs and units.[1]

**e.** Now that we have some confidence that the acceleration is constant in this experiment, take a look at the velocity-time and acceleration-time graphs. Were your predictions about the signs of the velocity and acceleration in Activity 6.3.1 correct? If not, explain what led to the predictions you made.

---

[1] If you didn't allow for the time offset when performing the fit, your "initial" position and velocity values will probably seem way off. Including the time offset leads to "initial" position and velocity values that correspond to the portion of the video you are analyzing. If you forgot to include the time offset in your fit, just perform the fit again.

**f.** It's worth returning to a persistent misconception that may have arisen in Activity 4.3.3. Based on your results, describe how the velocity and acceleration change throughout the ball's motion (on the way up, at the top, and on the way down). Use terms like positive, negative, speeding up, slowing down, constant, zero, etc.

---

When an object is dropped near the surface of Earth and the only force of consequence acting on it is the gravitational force, we say the object is in *free-fall*. For the experiment just performed, you should have found an acceleration that was close to (but probably not *exactly*) $-9.8\,\text{m/s}^2$. You might remember this value from a previous physics class as the "acceleration due to gravity." A few comments are in order regarding this value.

First, recall that the sign is related to the *direction* of the acceleration, which depends on the coordinate system we choose. The motion sensor defines the positive direction as pointing up in this experiment. Therefore, a negative acceleration tells us that the acceleration vector points *down*. Second, the measured acceleration may not be exactly what you expected. The reason for this disagreement is likely due to air resistance on the ball's motion; it can be surprisingly challenging to record free-fall data that is not affected by air resistance to some degree.[2] Third, the phrase "acceleration due to gravity" can lead to misconceptions about how the gravitational force acts on objects. A better phrase might be "free-fall acceleration," but even this has its problems. What we really need to understand is the gravitational *force* and how this force relates to the acceleration.

Wests/Getty images

## 6.4 THE GRAVITATIONAL FORCE

In the previous activity, we saw that a ball experiences an acceleration *without* the aid of a *visible*, applied force. However, if we believe Newton's second law to be true—*and we do!*—then the fact that the object accelerates implies that there *must* be a force acting on the object. We call this the *gravitational force*. More specifically, because this force seems to emanate from Earth, we call it the gravitational force of Earth acting on the ball. Our task in this section is to investigate this gravitational force. For now, we will only consider the gravitational force near the surface of Earth.

---

[2] Any object, including a basketball, will be affected by air resistance to some degree, but air resistance can be minimized by choosing a smaller, heavier object. Probably the best one can do in a classroom setting is to use a bowling ball. Of course, one must be extremely careful when tossing a bowling ball, as it can do serious damage if it lands on a motion sensor or someone's foot!

In order to carry out the activities in this section, you will need the following items:

- 1 balance or electronic scale
- 1 electronic force sensor (or spring scale)
- 1 set of hanging masses
- 1 hanging mass pan

### 6.4.1. Activity: Initial Thoughts About the Gravitational Force

**a.** Do you think the gravitational force is the same on all objects? Try to give a reason for your answer.

**b.** Do you think the gravitational force depends on the location of the object in the room? For example, would the gravitational force be different if the object were on the floor, on the table, or up near the ceiling? Again, try to give a reason for your answer.

Many people have different views on the above questions, so there may have been some disagreement within your group. But even if your group agreed on the answers, it is a relatively simple matter to investigate these questions experimentally. To do so, we will need to measure the gravitational force, which we will do with the help of an electronic force sensor. Remember, whenever you're using a force sensor, it is important to *zero the sensor in the correct orientation* before performing the experiment!

### 6.4.2. Activity: The Gravitational Force at Different Locations in the Room

**a.** Set up an experiment to measure force using a force sensor with an experiment time of around 60 seconds. Make sure the force sensor is set to the $\pm 10\,\mathrm{N}$ range, and zero the force sensor so it is oriented with the hook pointing down. Start the experiment and after about 5 seconds attach a 500-g mass to the sensor and hold it steady. After about 10 seconds, remove the mass for about 5 seconds and repeat the process so you have three separate force measurements. Select a several-second portion of each of these force measurements and report the averages below. (These averages may not be *exactly* the same, but they should be pretty close.)

**b.** The only thing pulling on the force sensor in part (a) is the hanging mass, and the reason the mass is pulling on the sensor is because of the gravitational force acting on it. Therefore, the force sensor reading tells us the magnitude of the gravitational force acting on the mass. We now want to repeat the experiment above, but this time while holding the mass down near the floor. (As always, be sure to zero the force sensor in the appropriate orientation before starting the experiment.) What values do you measure down near the floor? Are your measurements for the gravitational force consistent with what you found in the previous experiment?

**c.** Finally, let's repeat the experiment one more time, but this time hold the mass up near the ceiling (being sure that you're holding the force sensor in the correct orientation). What values do you measure up near the ceiling? Report your measurements below and explain whether the gravitational force appears to change at different locations in the room.

### The Gravitational Force Near Earth's Surface

You should have found that the gravitational force acting on an object remains the same regardless of where the object is located in the room. Technically, all we can really say is that the gravitational force is the same *within the uncertainty of our force sensor*. As it turns out, the gravitational force decreases as you move farther away from Earth's surface, but it only changes by about 0.0001% between the floor and the ceiling. Thus, for our purposes it is safe to assume the gravitational force is *constant* near the surface of Earth.[3]

Now that we know how to measure the gravitational force on an object, it is a relatively simple matter to determine if the gravitational force acting on different objects is the same. In the process, we will find a formula to calculate the gravitational force on any object near the surface of Earth.

### 6.4.3. Activity: A Formula for the Gravitational Force

**a.** Set up an experiment that runs for 120 seconds. In this experiment you will be hanging different masses from a force sensor, so be sure to zero the force sensor in the correct orientation. Begin the experiment, wait about 5 seconds, and then attach a hanging mass pan to the force sensor and hold it steady. After about 10 seconds add a 100-g mass to the

---

[3] The gravitational force is only about 1% weaker at an altitude of 30 km, which is about three times the cruising altitude of a large jet airplane.

pan and again hold it steady for about 10 seconds. Continue this process of adding masses until you have six or seven measurements. By taking appropriate averages, determine the magnitude of the gravitational force $F_g$ acting on the different masses and fill in the table below (don't forget to include the mass of the mass pan!). Describe your results. **Note**: It is more convenient to perform this experiment in *event mode*, if available.

| Mass (kg) | $F_g$ (N) |
|-----------|-----------|
|           |           |
|           |           |
|           |           |
|           |           |
|           |           |
|           |           |

**b.** Using a spreadsheet (or other appropriate software), make a graph of the gravitational force as a function of mass. Make a sketch of your graph. You should see that force and mass are proportional to each other. Fit a straight line to the data and determine the proportionality constant (be sure to include units).

**c.** The proportionality constant in this experiment tells us how much gravitational force is exerted per kilogram of mass. Because this constant is a direct result of Earth's gravitational force on an object, it is called the *gravitational field strength* and is given the symbol $g$. You now have everything you need to write down an equation (in symbols) that gives the magnitude of the gravitational force $F_g$ acting on an arbitrary object of mass $m$. Do this below.

## 6.5  ACCELERATION NEAR EARTH'S SURFACE

In the previous section, we found that the magnitude of the gravitational force is proportional to the mass of the object and is given by the simple equation

$$F_g = mg \tag{6.1}$$

where $m$ is the object's mass and the local gravitational field strength is given by the constant $g = 9.8$ N/kg. The direction of the gravitational force is toward the center of Earth, which we typically refer to as *down*.

Notice that the constant $g$ here has *nothing* to do with motion and, in particular, is *not* an acceleration! Do not make the mistake of confusing the above equation with Newton's second law. Equation (6.1) gives the magnitude of the gravitational force acting on an object of mass $m$ near the surface of Earth; it is merely one of the forces that may appear as part of the net force in Newton's second law.

As we learned at the end of the last unit, a newton is defined as the force needed to accelerate a 1 kg mass at a rate of $1\,\text{m/s}^2$, so that $1\,\text{N} = 1\frac{\text{kg}\cdot\text{m}}{\text{s}^2}$. Using this definition, we can deduce that 1 N/kg is equivalent to $1\,\text{m/s}^2$, which means it wouldn't be *incorrect* to write $g$ as $9.8\,\text{m/s}^2$. However, it should be clear from the previous experiment that the constant $g$ that appears in Eq. (6.1) has nothing to do with acceleration. As such, we will always use units of N/kg when using Eq. (6.1) to remind us that $g$ is the gravitational field strength and *not* an acceleration.[4] Of course, if we are calculating an *acceleration* and the units come out to be N/kg, then it is perfectly acceptable—in fact, we *urge* you—to convert these units into $\text{m/s}^2$.

Equation (6.1) tells us that the gravitational force is larger on objects that have a larger mass. In other words, the more massive an object, the larger the gravitational force. Our next task is to determine how this force affects an object's free-fall acceleration. In order to do the activities in this section, you will need the following items:

- 1 steel ball
- 1 rubber ball
- (OR two other objects with different masses that are relatively unaffected by air resistance)
- 1 ruler (with holes in it)

### 6.5.1. Activity: Free-Fall Acceleration

a. We began this unit by dropping a coffee filter and a ball, two objects with obviously different masses, at the same time. For the situation when *air resistance was negligible* (using the crumpled coffee filter), recall that both objects hit the ground at nearly the same time. What do you think you will observe if you drop a steel ball and a rubber ball (or a paperclip and a pencil) at the same time?

b. Let's perform this experiment using a steel ball and a rubber ball (or a paperclip and a pencil). To release the two objects at the same time, it can be helpful to set them both on a ruler with holes in it (for the balls) or a hardcover book (for the paperclip and pencil). Hold the ruler (or book)

---

[4] You might find the fact that the constant $g$ can be written with two different sets of units to be confusing. This subtlety is a direct result of the fact that the gravitational mass and the inertial mass are both measured in kg, even though they refer to different physical properties.

with your arms straight out in front of you at approximately eye level, being careful to hold it steady and level. Then, when you are ready, push the ruler (or book) *straight down* as quickly as you can, allowing the two objects to drop. (It's important to move the ruler *straight down* without first moving up or twisting at all; this might take a little practice.) If you perform the maneuver correctly, your head will naturally follow the objects down as they fall, which makes it easy to see what's happening. What do you observe? Does one hit the ground first, or do they hit at roughly the same time? Does this match your prediction? (If you find it difficult to see what's happening, try listening to hear whether there is a single sound or two separate sounds as the objects hit the ground.)

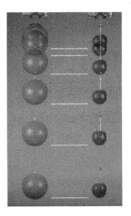

If you perform the above experiment carefully, it should be clear that both objects hit the ground at the same time. In the absence of air resistance, any two objects dropped at the same time will fall in exactly the same way and will hit the ground at the same time. Of course, because air resistance is almost always at play in the real world, our everyday experience suggests that a heavier object will land *slightly* before a lighter object. But careful experiments done in controlled situations demonstrate very clearly, and perhaps surprisingly, that objects with different masses accelerate at exactly the same rate.[5] Let's see if we can understand this observation using Newton's second law. To do so, we need to introduce an important concept known as a *free-body diagram*.

### The Free-Body Diagram: An Introduction

In its basic form, a free-body diagram is constructed by drawing a simple representation of an object, perhaps a circle to represent a ball or a rectangle to represent a block of wood, and then drawing an arrow to represent each force acting on the object. The arrows are drawn according to the following rules:

- The tail of each arrow is placed at the point where the force is acting.
- Each arrow points in the direction the force acts and is labeled by its magnitude.
- The length of each arrow is proportional to the magnitude of the force.

There are a few important details worth mentioning. First, the point at which a force acts is not always obvious, and some forces act at more than one location. For example, there is a tiny gravitational attraction between every point in Earth and every point in an object. However, the result of all these micro-forces is extremely well approximated by a single force acting at the *center-of-mass* of

---

[5] A truly spectacular example of free-fall with no air resistance is demonstrated in an episode of the BBC series *The Human Universe*, where a bowling ball and some feathers are dropped from a height of over 100 feet in a near perfect vacuum. This particular scene can be viewed on the BBC's YouTube channel at https://www.youtube.com/watch?v=E43-CfukEgs.

the object and pointing to the *center-of-mass* of Earth.[6] We will discuss how to handle such forces as we encounter them.

Second, we often do not know the magnitude of a particular force in advance. In such a situation, the force vector (arrow) is labeled using the magnitude notation previously discussed. For example, if the gravitational force is represented by the vector $\vec{F}_g$, then the magnitude of the gravitational force is written as $F_g$ (without the arrow). As we now know, the magnitude of the gravitational force is given by $F_g = mg$, where $m$ is the object's mass and $g = 9.8$ N/kg is the gravitational field strength. We can therefore label the gravitational force vector on a free-body diagram as either $F_g$ or $mg$.

Lastly, it is not always clear when we have accounted for all the forces. How can we be sure we haven't missed a force in our free-body diagram? To help answer this question, we distinguish between *contact* forces and *non-contact* forces. Contact forces result from physical contact with an object. Anything that touches an object—a table, a string, your hand, the air, etc.—can exert a force on the object. Conversely, non-contact forces, such as the gravitational force, are a result of "action at a distance," meaning there is nothing in physical contact with the object.

Because contact forces arise from physical contact, they are relatively easy to spot: you simply need to look and see what's *touching* the object.[7] Non-contact forces, on the other hand, are a bit trickier. Fortunately, there are very few non-contact forces to worry about. Apart from the gravitational force, the only other non-contact forces we will encounter are electric and magnetic forces. Thus, the procedure for identifying forces is to first account for the non-contact forces, which is often only the gravitational force, and then look to see what's touching the object. Generally speaking, anything touching the object will give rise to at least one contact force (some of which may end up being neglected).

Once we've drawn a free-body diagram, the next steps are to choose a coordinate system and apply Newton's second law in component form. This procedure is outlined in the following activity.

### 6.5.2. Activity: Acceleration of a Dropped Object

**a.** As in the previous activity, consider a ball of mass $m$ that's been dropped and is falling to the ground. Our first step is to draw a free-body diagram of the ball after it has been dropped and before it hits the ground. Thus, we need to find all the forces that are acting. Begin by specifying any non-contact forces, and then describe anything touching the object (careful, there *is* something touching the object).

---

[6] We will discuss an object's center-of-mass in a later unit. For now, you can assume the center-of-mass is located at the center of the object.

[7] It should be noted that because air is invisible it can be easy to overlook any forces exerted by the air. Fortunately, air forces are usually unimportant and are therefore neglected in most situations.

**b.** In this problem, the only non-contact force is the gravitational force, and the only thing touching the object (after it has left our hand) is the air. If we neglect air resistance, we end up with only a single force. Below, draw a free-body diagram of the ball for this situation (don't forget to label the force arrow with its magnitude). Next to your free-body diagram, sketch a standard coordinate system ($x$-axis pointing to the right and $y$-axis pointing up).

**c.** Using this coordinate system, apply Newton's second law in the $y$-direction. That is, sum up the force components in the $y$-direction (there's only one) and set it equal to $ma_y$ (be mindful of signs). Note that we are assuming we do *not* know $a_y$ here; this is what we are trying to find!

**d.** Finally, solve the resulting equation for $a_y$ and substitute in the numerical value for $g$. What do you find for the acceleration? Does your answer depend on the mass $m$ of the object?

---

You should have found an acceleration of $a_y = -9.8$ m/s², which is *independent* of the mass $m$ (consistent with our observations from Activity 6.5.1). In other words, all objects will fall with the same acceleration in the absence of air resistance. We say that objects near the surface of Earth have a *free-fall acceleration* of 9.8 m/s². (If you're being careful with units, you should have found that $a_y = -9.8$ N/kg, but because we are calculating an acceleration, it is appropriate to convert the units from N/kg to m/s².) The minus sign in our answer is a result of our choice of coordinate system and tells us that the acceleration points in the negative $y$-direction (down). This result is only strictly true when there's *no* air resistance; for many objects, this will not be the case, as the following activity makes clear.

### 6.5.3. Activity: Air Drag

**a.** Using the same motion-sensor setup as in Activity 6.3.2, track the motion of a dropped (uncrumpled) coffee filter. Because the motion of the coffee filter is strongly affected by air resistance, it may float to the left or right after being dropped. Thus, you'll probably need to try the experiment a few times before you are able to track its motion. Once you have clean data, make a sketch of the velocity-time graph (you can ignore any small bumps in your actual data; just try to sketch the basic shape of the graph).

**b.** Notice that the coffee filter reaches a constant speed after a short period of time. We call this the object's *terminal speed*. What is the (approximate) terminal speed of the coffee filter? What is the coffee filter's acceleration once it reaches its terminal speed?

**c.** Below, draw a free-body diagram of the coffee filter (mass $m$) as it is falling to the ground; include a standard coordinate system next to your diagram. For this situation, in addition to the gravitational force, there is a *drag* force from the air that cannot be ignored. This drag force points in the opposite direction of the motion and its magnitude changes with time (the magnitude actually depends on the object's speed). Label this force $F_{drag}$.[8]

**d.** Now apply Newton's second law in the $y$-direction. That is, sum up the force components in the $y$-direction and set it equal to $ma_y$ (be mindful of signs). Solve for the magnitude of the drag force $F_{drag}$.

**e.** Once the object reaches terminal speed, it's acceleration will be zero. Use this information to determine the magnitude of the drag force when the filter is at its terminal speed.

---

[8] Although the drag force acts over the entire bottom surface of the coffee filter, we can draw it as a single force that acts at the center of the filter.

The magnitude of the drag force depends on an object's speed. You may have experienced this when traveling in a car and sticking your arm out the window. If the car is moving slowly you won't feel much of a force, but if the car is traveling at freeway speeds the force on your arm can be quite significant. Because of this speed dependence, an object dropped from rest will initially experience a very small drag force. But as the object speeds up, the drag force increases, which results in a smaller acceleration. This process continues until the object reaches its terminal speed, at which point the drag force is equal to the gravitational force (in magnitude) and the acceleration becomes zero.

It's worth pointing out a significant difference between Activities 6.5.2 and 6.5.3. In Activity 6.5.2, we knew the forces but didn't know the acceleration, whereas in Activity 6.5.3, we knew the acceleration (once the object reached terminal speed) but didn't know one of the forces. These are essentially the only two situations that can arise when using Newton's second law to solve problems. Thus, one of the key questions to ask when using Newton's second law is whether you know all the forces (or can figure them out), or whether you know what the acceleration is.

### Mass, Weight, and Gravitational Force

It is easy to confuse the concepts of mass, weight, and gravitational force. Mass is an *intrinsic property* of an object and can be casually defined as the "amount of stuff" comprising it. The gravitational force, on the other hand, is a force that one object (such as Earth) exerts on another. Finally, *weight* is a quantity that is determined by putting an object on a scale when there are *no other forces present* (besides the gravitational force). For example, to determine your body weight, you stand on a scale, the gravitational force of Earth pulls down on you, and this downward force causes the scale to read a particular value. Like the gravitational force, weight is measured in newtons (or pounds), but unlike a force weight does not have a direction.[9] In this sense, weight can be thought of as the magnitude of the gravitational force.

Aabejon/Getty images

One potentially confusing aspect of weight is that additional forces can be present that appear to change the weight of an object. In such situations, you may hear the term *apparent weight*. For example, when you get into a pool of water, you have likely experienced a sense of near "weightlessness." The gravitational force acting on you has not changed, but the water now exerts an upward force on you (the *buoyant force*). The result is that the net force downward is much smaller than the force of gravity, and this leads to an apparent weight that is much smaller than your actual weight. If you put a scale on the bottom of a pool and stood on it, the reading would be much smaller than usual, even though your actual weight, as we have defined it, has not changed.

To add to the confusion, consider an astronaut standing on the moon. Their weight as measured on the moon will be different from their weight as measured on Earth because the gravitational force on the moon is different from the gravitational force on Earth ($g_{\text{moon}} \approx g_{\text{Earth}}/6 = 1.6$ N/kg). In a situation like this, we would need to specify their "weight on Earth" or their "weight on the

---

[9] It is worth mentioning that many textbooks define weight to be the gravitational force of Earth acting on the object, making it a vector.

moon." Even more confusing is when one is in orbit around Earth, such as on the International Space Station. In this situation, the gravitational force is still quite strong, but the astronauts experience weightlessness.

### 6.5.4. Activity: Mass and Weight

a. We have said that mass is a measure of the "amount of stuff" in an object. Based on this statement, compare the *mass* of the astronaut in three different locations: on Earth, in orbit on the space station, and on the surface of the moon. Briefly explain.

b. We have said that weight is the result of a measurement of putting an object on a scale with no other forces present. If you have seen videos of astronauts on the space station, you probably remember that they (and anything not strapped down) appear to "float" around the station. Based on this observation and our definition of weight, compare the *weight* of the astronaut in the three different locations: on Earth, in orbit on the space station, and on the surface of the moon. Briefly explain.

c. Finally, compare the *gravitational force* on the astronaut in the three different locations: on Earth, in orbit on the space station, and on the surface of the moon. When considering the astronaut on the space station, you will need to use the fact that the local gravitational strength at the orbital distance of the space station is given by $g_{\text{space station}} = g/1.15 = 8.5$ N/kg.[10]

It may come as a surprise to learn that the gravitational force on an astronaut in the space station is nearly as large as the gravitational force on the surface of Earth. Nevertheless, the astronaut in the space station is weightless, at least by our definition. The fact that the astronaut is weightless on the space station is not because the gravitational force of Earth is zero at their location. Instead, when in orbit, the astronaut (and the entire space station) is continuously in free-fall toward the center of Earth. However, their velocity in the

---

[10] Later in this unit, we will find out why the gravitational strength at the location of the space station is lower than on the surface of Earth.

horizontal direction is so large that the surface of Earth curves away from them as they fall.[11] We will return to discuss this idea in more detail at the end of this unit.

# MOTION IN MORE THAN ONE DIMENSION

## 6.6  MOTION WITH A CONTINUOUS FORCE

Earlier, we found that near the surface of Earth the gravitational force is constant and proportional to the mass of an object:

$$F_g = mg$$

where the constant $g = 9.8$ N/kg is the gravitational field strength, and the direction of this force is downward (toward the center of Earth). This equation tells us that objects with larger mass experience a larger downward force due to gravity. You experience this any time you lift something up; to lift an object you need to overcome the gravitational force pulling it down. A feather is obviously much easier to lift than a book, which itself is much easier to lift than a car. This is simply due to the fact that the gravitational force increases with mass.

But even though the gravitational *force* depends on an object's mass, we saw that two objects of different mass fall in exactly the same manner in the absence of other forces like air resistance. Specifically, the *acceleration* they experience is the same (and has a magnitude of $9.8$ m/s$^2$) even though the forces are different. This surprising behavior is due to the (gravitational) mass dependence of the gravitational force canceling with the (inertial) mass that appears in Newton's second law.

It's worth noting that the gravitational force, which acts vertically, has no effect on our horizontal motion experiments like the cart on a track.[12] In other words, motions in the horizontal and vertical directions are independent of one another. We will observe this direction independence shortly, when we investigate the motion of an object that is launched near the surface of Earth. Such motion is commonly referred to as *projectile motion*.

Before studying motion in two dimensions, we would like to do an experiment that demonstrates the physical nature of a constant, continuous force in one dimension. For example, a dropped ball experiences a constant downward force due to gravity. We can create a horizontal analog of this motion by continuously tapping a ball in one direction on a flat surface. The similarity between a falling ball and a tapped ball will help us study projectile motion, in which an object falls vertically while simultaneously moving horizontally.

For the measurements described below, you will use a twirling baton with a rubber tip to tap a bowling ball and watch its motion. You should have the following equipment available:

- 1 bowling ball (or other heavy ball)
- 1 twirling baton (or other stick used to tap the ball)

---

[11] Clearly, Earth cannot be seen as flat from their perspective.

[12] We did need to be careful to keep the motion only in the horizontal direction—any "tilt" to the cart track would have led to an acceleration "downhill" caused by the gravitational force.

Find a stretch of smooth, level floor over which the ball can roll for some distance (a hallway can be used if the classroom is not large enough). Over the next three activities you will watch the motion of the ball for different situations: (1) a briskly rolling ball receiving no taps, (2) a ball starting at rest and receiving regular taps, and (3) a ball with an *initial* velocity along one direction and receiving regular taps in the direction *opposite* to its initial velocity ("tapped backwards"). For each situation, make a qualitative observation of the motion. For example, does the ball speed up? Slow down? Keep rolling at roughly the same speed?

### 6.6.1. Activity: The Motion of a Freely Rolling Ball

**a.** Assume that the ball is rolling freely and without friction so there is no net force acting on it. What do you predict the nature of the motion will be? Will the velocity change? Will there be an acceleration? Sketch *predicted* graphs of position vs time and velocity vs time.

**b.** Now perform the experiment. Give the ball a push to get it started and observe its motion after your push. What happens to the velocity? Is there an acceleration? Does the observed motion match your prediction? (Your instructor may ask you to make quantitative measurements of the ball's motion by, for example, dropping bean bags at the location of the ball each second. If so, sketch your measured position-time graph.)

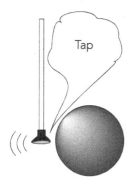

Fig. 6.3. When a bowling ball receives a series of taps, how does it move?

You should have seen that the ball continues rolling at a (roughly) constant velocity in the absence of any applied forces (the acceleration is zero). The position as a function of time would be a *linear* graph with a constant slope, while the velocity graph would be a horizontal line with zero slope (but non-zero value). Not surprisingly, this is the same motion we observed for a low-friction cart rolling with no applied force.

### 6.6.2. Activity: 1D Ball Tapping Starting from Rest

**a.** Now consider a ball that starts at rest and receives a series of steady taps in one direction (Fig. 6.3). What do you predict the motion of the ball will be? Will the velocity change? Will there be an acceleration? Sketch *predicted* graphs of position vs time and velocity vs time.

**b.** Now perform the experiment. Start the ball at rest and begin tapping on one side of the ball using a series of quick, regular taps. Be sure to keep tapping at the same rate (and with the same strength) even as the ball starts to roll. What do you observe? Does there appear to be an acceleration? Does the observed motion match your prediction? (Once again, if you make quantitative measurements of the ball's motion, sketch your measured position-time graph.)

In this case, you should have seen that the ball accelerates at a (roughly) constant rate so that the velocity is continually increasing. The velocity as a function of time would be a *linear* graph, while the position as a function of time would be a *quadratic* (parabolic) graph. Note that this motion is analogous to a dropped ball: from an initial velocity of zero, a dropped ball continually speeds up in the downward direction. Your continuous tapping on the ball serves as the applied force instead of the gravitational force.

### 6.6.3. Activity: 1D Ball Tapping with Initial Motion

**a.** Lastly, we examine the situation in which the ball has an initial velocity in a direction that is opposite to the direction of the applied force; you will start the ball rolling in one direction, and then give it a series of quick, regular taps in the *opposite* direction (against the motion). What do you predict the motion of the ball will be? Will the velocity change? Will there be an acceleration? Sketch *predicted* graphs of position vs time and velocity vs time.

**b.** Now perform the experiment. Start the ball rolling in one direction and then begin tapping on the ball using a series of quick, regular taps that exert a force in the *opposite* direction of its motion. Keep tapping until the ball eventually stops, turns around, and starts moving back in the direction it came. What do you observe? Does there appear to be an acceleration? Does the observed motion match your prediction? (If you make quantitative measurements, sketch your measured position-time graph.)

This time the ball has an initial velocity in one direction, but the continuous tapping force creates a (roughly) constant acceleration in the opposite direction. The velocity starts out with a large magnitude, then decreases to zero before eventually increasing again (in the opposite direction). The velocity as a function of time would again be a *linear* graph, but this time it would cross through zero. The position as a function of time would again be a *quadratic* (parabolic) graph, but this time the curve will exhibit a maximum or minimum as the ball turns around. Note that this motion is analogous to motion we have seen before: a ball being tossed in the air!

The next ball-tapping experiment is more challenging. In this activity, you roll the ball briskly in one direction and then, while it is rolling, tap on it *perpendicular to its original direction* (not perpendicular to its current direction). For example, if you initially start the ball rolling down a hallway (call this the *x*-direction), you will continuously tap it toward one of the side walls (call this the negative *y*-direction).

---

### 6.6.4.  Activity: 2D Ball Tapping with Initial Motion

**a.** On the graph below try to *predict* what the resulting graph of the ball's two-dimensional motion will look like. Use your previous observations of the different types of motion we have seen. In particular, what is the acceleration in the *x*-direction? In the *y*-direction? Explain the basis for your prediction.

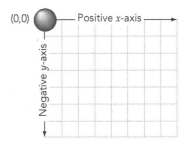

**b.** Now try the experiment. Start a ball rolling in one direction along the floor, and while it is rolling tap on it *perpendicular to its original direction*. Does the observed motion agree (approximately) with your prediction? In particular, do the *x* and *y* motions appear to be independent? How can you tell? (If you make quantitative measurements, sketch your observed trajectory.)

## 6.7 PROJECTILE MOTION

Thus far, we have *separately* dealt with one-dimensional horizontal motion (cart on a track or tapping a ball) and one-dimensional vertical motion (dropped ball or a ball tossed straight up). Of course, most types of motion are more complicated, such as when you toss an object to a friend, or the (nearly) circular orbit of Earth around the sun.

Imagine throwing a rock off a cliff with an initial velocity at an angle $\theta$ with the horizontal (see Fig. 6.4). As you may know, the rock will continue to move forward in the horizontal direction, only slowing down a small amount due to air resistance. If there were no gravitational force, the rock would continue in a straight path, up and away from you. But due to the gravitational force between the rock and Earth, the rock experiences an acceleration in the (negative) vertical direction. In other words, the rock starts to "fall." Taken together, the full motion is two-dimensional, and we will need to keep track of what happens in both the $x$ and $y$ directions. This type of motion is known as *projectile motion.*

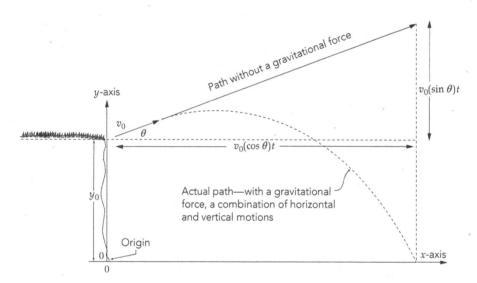

**Fig. 6.4.** Diagram of resulting motion with and without the gravitational force.

In the following activity, we analyze the motion of a basketball shot, a common example of projectile motion where the constant force in one dimension is the gravitational force. Your instructor may provide additional information depending on the program (or movie) you are using.

### 6.7.1. Activity: Basketball Shot

**a.** Open the movie <Basketball Shot> in your video analysis program. Watch the movie a few times, and then *predict* what the position-time graphs look like for both the horizontal and vertical motions. Assume a standard coordinate system and limit your prediction to the time *after* the ball leaves the person's hand until just *before* it bounces. Draw your predictions for $x(t)$ and $y(t)$ as dashed lines on separate graphs and explain briefly why you sketched them the way you did. Note that we are *not* asking you to predict the actual trajectory $y(x)$.

**b.** After scaling the movie carry out the analysis by clicking on the center of the ball to track its position (you should only track the motion after the ball has left the person's hand until just before it bounces). As you click on the basketball you should see position-time data plotted on a graph. How do these graphs compare to your predictions? Sketch the actual results on your graphs in part (a) using solid lines. **Note**: Your instructor may ask you to open a file in which the analysis has already been performed.

**c.** Use the software to separately *fit* the data for both $x(t)$ and $y(t)$, choosing appropriate functions for each. **Reminder**: Only fit the portion of the graph *after* the ball leaves the person's hand until just *before* it bounces. Also, be sure to enable the time offset when performing the fit (since the ball doesn't leave the person's hand at exactly $t = 0$). Write down the two fit equations below, including the values of the constants with appropriate units.

**d.** Note that your fit equations are simply the kinematic equations for the horizontal and vertical position. Using these equations, identify the following values: $v_{0x}$, $v_{0y}$, $a_x$, and $a_y$, and use them to write down the kinematic equations for $v_x(t)$ and $v_y(t)$. (You should be able to write down these equations without making any new graphs!)

**e.** Based on the observed motion and the equations you found, sketch the following graphs: $v_x(t)$, $v_y(t)$, $a_x(t)$, and $a_y(t)$.

An important result here is that the horizontal and vertical motions are independent of each other. In other words, the force in the vertical direction does not affect what happens in the horizontal direction, and vice versa. Ultimately, this behavior is a result of the fact that position, velocity, acceleration, and force are *vector* quantities, implying Newton's second law can be separated into three component equations. The following activity provides a nice demonstration of this independence of horizontal and vertical motion.

### Independence of Horizontal and Vertical Motions

Consider the situation shown in Fig. 6.5. Someone riding on a moving cart with negligible friction tosses a ball vertically into the air. Where will the ball land when it returns to ground level? Does it depend on how fast the cart is moving?

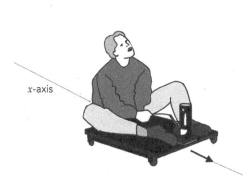

**Fig. 6.5.** Diagram of a person launching a ball straight up while moving at a constant horizontal velocity.

For this demonstration you will need:

- 1 kinesthetic cart
- 1 spring-loaded projectile launcher with ball

---

### 6.7.2. Activity: The Path of a Ball Launched Vertically from a Moving Cart

**a.** Describe what you think will happen when a ball is launched vertically from a moving cart. Sketch the path of the ball as an observer at rest in the laboratory would see it (i.e., what will you see from your chair as the rider moves by?). Explain briefly.

**b.** Where do you think the ball will land when it comes back down? In front of the rider? Behind the rider? Right back at the launcher? Explain briefly.

**c.** Now observe the actual demonstration and describe what happens. How does it compare to your predictions?

**d.** To make this more concrete, let's put in some numbers. Suppose the cart is moving at a constant speed of 2 m/s along the ground when the rider launches the ball. What is the ball's initial *horizontal* velocity component according to an observer on the ground? What is its *horizontal* velocity component when it's at its highest point (assuming air resistance can be ignored)? How about when it returns to floor level?

**e.** Suppose the ball is launched vertically with a speed of 2 m/s. Describe the vertical motion of the ball as it moves through its trajectory.

---

This demonstration provides a nice example of inertial reference frames and the fact that the laws of physics are the same in all inertial frames. Recall that an inertial reference frame is any frame of reference that is not accelerating. Thus, the rider on the cart and a person on the ground each represent a different inertial reference frame. According to the rider on the cart, the ball is launched straight up, and if air resistance is negligible, the only force acting is the gravitational force (which acts vertically). Therefore, the rider will predict that the ball should go straight up and come straight back down, landing on (or near) the launcher.

Meanwhile, an observer on the ground will see the cart (and everything on it, including the ball) moving with a horizontal velocity. Assuming air resistance is negligible (and assuming the cart maintains a constant velocity), once the ball is launched the only force acting is the gravitational force (which acts vertically). Therefore, a ground observer will predict that the horizontal velocity component will remain constant while the ball travels up and comes back down, thus traveling in a curved trajectory that ends up landing on (or near) the launcher. Although the two observers predict different *trajectories* due to their different reference frames, they can both use Newton's second law to analyze the motion, and they both predict the ball should end up back at the launcher (which it does).

### 6.8 PROBLEM SOLVING

#### 6.8.1. Activity: Projectile Range with a Launch Angle

A batter hits a pitched ball when the ball is at a distance of 1.2 m above the ground, and the ball leaves the bat with an unknown speed at an angle of 40° above the ground (Fig. 6.6). The ball is observed to have a *range* of 107 m; in other words, the ball will hit the ground 107 m *in the horizontal direction* away from the batter (assuming no one catches it and that it doesn't hit the outfield fence). Although it's not fully justified in this situation, air resistance should be ignored.

Fig. 6.6. A batted ball in a softball game. Image credit: motionshooter/ Adobe stock.

**a.** Use the range to determine the *speed* with which the ball leaves the bat. (We want the total speed here, not just one of its components.)

**b.** Assume the outfield fence has a height of 3.3 m and is 100 m away from the batter. Does the ball make it over the fence? If so, how high above the fence does the ball pass? If not, where will the ball hit the fence (e.g., how far *below* the top of the fence)?

## ORBITAL MOTION AND UNIVERSAL GRAVITATION (OPTIONAL)

### 6.9   CONNECTING PROJECTILE MOTION TO ORBITAL MOTION

Now that we have the tools to handle two-dimensional projectile motion, let's look at the situation of throwing a ball horizontally. While the analysis is relatively simple, the result can be extended without too much difficulty to help us understand the phenomenon of orbital motion.

#### Throwing a Ball

Suppose you throw a ball horizontally as hard as you can, and you want to figure out how far it will travel. Because people have different sizes and abilities, we will assume the ball is thrown from a specific (but arbitrary) height $h$ with a specific (but arbitrary) initial speed $v_0$, assumed to be entirely in the horizontal direction, as shown in Fig. 6.7. We will ignore air resistance throughout this analysis.

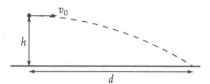

Fig. 6.7. A sketch of the trajectory of a horizontally thrown ball.

---

#### 6.9.1.   Activity: The Range of a Horizontally Thrown Ball

**a.** We want to determine the horizontal distance $d$ traveled by a ball of mass $m$ that is thrown horizontally with speed $v_0$ from an initial height $h$. Let's begin by defining a coordinate system with the $x$-axis pointing horizontally (in the direction the ball is thrown) and the $y$-axis pointing vertically up, with the origin at ground level underneath the point where the ball is released. Given this coordinate system, determine and write down the initial positions ($x_0$ and $y_0$) and velocities ($v_{0x}$ and $v_{0y}$) of the ball in terms of the problem variables.

**b.** Once the ball has been released, the only force acting on the ball is the gravitational force (assuming air resistance is neglected). Draw a free-body diagram of the ball and determine the net force components $F_x$ and $F_y$.

**c.** Write down Newton's second law in both the $x$ and $y$ directions and determine the acceleration components $a_x$ and $a_y$.

**d.** You should find that both acceleration components are constant, which means we can use the kinematic equations in both the horizontal and vertical directions. Use the kinematic equations to solve for the horizontal distance $d$ traveled by the ball when it hits the ground. Your answer should be written in terms of the parameters $v_0$, $h$, and $g$. (This will probably take a couple of steps, so be sure to show your work.)

**e.** Next, use your result from part (d) to determine the *minimum* speed needed to throw a ball horizontally from the pitcher's mound to home plate on a major league baseball field (a horizontal distance of 18.44 m). You can assume the ball is thrown from a height of 2.0 m and hits the ground right at home plate. Convert your answer to miles per hour.

**f.** Do you think it would be possible, at least theoretically, for a ball to be launched with a large enough horizontal velocity to travel all the way around the planet (assuming, of course, that there is no air resistance, and the planet is completely smooth with no trees or buildings to get in the way)? Explain why or why not.

Hopefully, you did not have too much trouble deriving the formula $d = v_0\sqrt{2h/g}$, which gives the horizontal range for an object launched horizontally with no air resistance. In the next activity, we will use this formula to determine the speed needed to launch an object into orbit around Earth.

**Curvature of Earth's Surface**

Imagine viewing Earth from a large enough distance so that you can see the surface is curved. Assuming Earth to be perfectly spherical (and smooth, with no mountains, buildings, or trees), the outer edge of the surface will have a circular shape. Because of this shape, the surface will curve down instead of being perfectly horizontal, as shown in Fig. 6.8.

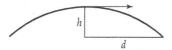

**Fig. 6.8.** Because Earth is spherical, it's surface curves down very slightly from a perfectly horizontal line.

As shown in Fig. 6.7, the trajectory of the thrown ball is also curved and forms the shape of a parabola. It turns out that the portion of a parabola near its apex looks an awful lot like a circle. In fact, it seems like we might be able to match up the curve in Fig. 6.7 with the curve in Fig. 6.8 so that the ball's trajectory would be aligned with the surface of Earth. The following activity pursues this idea in an attempt to calculate the speed needed to cause a horizontally launched ball to travel all the way around the planet.

---

**6.9.2. Activity: The Speed Needed to Orbit Earth at Its Surface**

**a.** To match the trajectory of the thrown ball to Earth's surface, we need to match the launching point and the landing point to two points on the surface and then determine how $h$ and $d$ are related to Earth's radius $R_E$. Thus, we assume the ball is launched *just above* the surface and travels a horizontal distance $d$ while descending vertically by a height $h$ and following the curvature of Earth, as shown in Fig. 6.8. Using the figure below as a guide, determine both $h$ and $d$ in terms of $R_E$ and $\theta$. **Hint**: Use trigonometry and think carefully about how to find $h$.

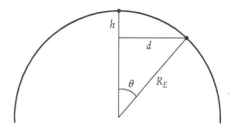

**b.** As mentioned, the parabolic trajectory of the ball and the circular shape of Earth are only approximately the same shape when the angle $\theta$ is small (*small* here means that $\theta \ll 1$ when $\theta$ is measured in radians). To proceed we need to approximate $\theta$ as a small angle. Two small-angle approximations quite common in physics are $\sin \theta \approx \theta$ and $\cos \theta \approx 1 - \frac{1}{2}\theta^2$. These approximations can be derived using a Taylor series, but for our purposes we just want to verify that these approximations are indeed valid. Using your calculator (be sure it is set to *radians* for angles), fill in the table below to verify that $\sin \theta \approx \theta$ and $\cos \theta \approx 1 - \frac{1}{2}\theta^2$ for small angles.

| $\theta$ (rad) | $\sin \theta$ | $\cos \theta$ | $1 - \frac{1}{2}\theta^2$ |
|---|---|---|---|
| 0.00 | | | |
| 0.05 | | | |
| 0.10 | | | |
| 0.15 | | | |
| 0.20 | | | |
| 0.25 | | | |
| 0.30 | | | |

c. Now that we have verified the approximations given above, replace $\sin \theta$ and $\cos \theta$ with their small-angle approximations in your expressions for both $h$ and $d$ from part (a).

d. All that remains is to substitute the small-angle expressions for $h$ and $d$ into the projectile range formula $d = v_0 \sqrt{2h/g}$ and then solve the resulting equation for the launch speed $v_0$. Carry out these steps and show that you end up with $v_0 = \sqrt{gR_E}$.

e. Using the known values for $R_E$ and $g$, determine the speed needed to launch an object into orbit just above the surface of Earth.

---

We hope you are impressed! Beginning with a relatively simple projectile motion problem and matching the trajectory to the curvature of Earth, we were able to determine the speed needed for an object to be placed into orbit (near the surface). While this situation is clearly not realistic—we can't neglect air drag and there are mountains, buildings, and trees that would get in the way—the calculation is theoretically sound. Moreover, it provides a good physical basis for what is happening when an object orbits a planet. In particular, the object is continually being accelerated toward the center of Earth, but it has such a large horizontal speed that by the time it falls to the point where it would have hit the ground, the surface has curved away so that the object is no closer to the ground than when it started. In other words, the object is in continual free-fall but gets no closer to the ground because the curved trajectory of the motion is the same as the curved surface of Earth.

At this point, you might have a sense of déjà vu. Way back in Unit 1 we used Kepler's third law to perform calculations on how the orbital period was related to the orbital radius. We have now come nearly full circle and are in a position to re-visit this idea.

### 6.9.3. Activity: The Orbital Period

a. Recall that the orbital period $T$ is the time it takes to complete one full orbit. Thus, traveling at the velocity you calculated above, the object will travel once around the planet in a time equal to one orbital period $T$. Use this information to find an equation for the orbital period $T$ in terms of $R_E$ and $g$ (leave everything in terms of variables). **Hint**: Find the distance traveled in one orbit and divide by the speed.

b. Plug in the known values for $R_E$ and $g$ to determine the orbital period for an object that is orbiting *just above* the surface of Earth. How does this compare to the orbital period of the space shuttle (see Activity 1.4.2)?

Unfortunately, we are not yet able to derive Kepler's third law. The issue is that we only have an expression for the gravitational field strength at the surface of Earth. To proceed, we need an expression for the gravitational field strength at an arbitrary distance from Earth. This is the topic of the next section.

## 6.10   UNIVERSAL GRAVITATION

Earlier, we measured the gravitational force on an object near the floor, at table height, and near the ceiling, and found that the gravitational force was constant (within experimental uncertainties) over this range. But as you are likely aware, the gravitational force *does* get smaller as you move farther away from Earth. The reason we couldn't measure this effect is because the force changes *very* gradually; you need to move very far from the surface before measuring any significant change in the gravitational force.

According to *Newton's law of universal gravitation*,[13] any two (spherical) massive objects are attracted directly toward one another with a force of magnitude

$$F_g = G\frac{m_1 m_2}{r^2}$$

---

[13] Newton deduced the law of universal gravitation from Kepler's third law, which was obtained by analyzing the observational data of Tycho Brahe (who made incredibly accurate naked-eye observations of the planets for over 20 years).

where $m_1$ and $m_2$ are the masses of the objects (in kilograms), $r$ is the distance between the centers of the two objects (in meters), and $G = 6.67 \times 10^{-11}$ N m$^2$/kg$^2$ is the gravitational constant. It is difficult to carry out an experiment to verify this formula because gravitational forces, in general, are extremely weak (owing to the tiny value of the gravitational constant $G$). In fact, gravitational forces between everyday objects can (essentially) always be neglected; gravitational forces only need to be considered when dealing with masses that are *very* large, such as the mass of a planet.[14]

### 6.10.1. Activity: The Strength of the Gravitational Force

**a.** Using Newton's law of universal gravitation, calculate the gravitational force between two 1-kg masses whose centers are separated by 10 cm.

**b.** *Estimate* the gravitational force between a 1-kg mass and Mt. Everest. To do this estimation, you can assume that Mt. Everest has a mass of 800 trillion kg and can be approximated as a sphere of radius 4,000 m, with the 1-kg mass sitting on the surface of this sphere.

**c.** Calculate the gravitational force between a 1-kg mass and Earth, with the mass sitting on the surface of Earth. (Earth has a mass of $M_E = 5.97 \times 10^{24}$ kg and a radius $R_E = 6.37 \times 10^6$ m.)

**d.** Now calculate the gravitational force acting on a 1-kg mass at the surface of Earth the "easy way," using $F_g = mg$. What do your answers to parts (c) and (d) suggest about these two gravitational formulas?

You should have noticed that the gravitational force calculated in parts (c) and (d) have the same value. Perhaps this is not surprising; after all, we were calculating the same thing using two different formulas. However, the fact that these two

---

[14] It is somewhat ironic that while gravity is the weakest of all forces, it is also the force that is most noticeable to humans. We will come back to this idea when we consider electric forces.

formulas give the same result suggests that we can determine a new formula for the gravitational field strength $g$.

### 6.10.2. Activity: The Gravitational Field Strength

**a.** In the previous activity, we found that the gravitational force on a 1-kg mass was the same whether it was calculated using Newton's law of universal gravitation or using the simpler formula in terms of the local gravitational field strength $g$. Below, write down the gravitational force acting on an *arbitrary* mass $m$ sitting on the surface of Earth using both formulas. Note that we want you to write down these formulas in terms of *symbols*, without any numbers.

**b.** These two formulas refer to the same thing—the gravitational force acting on a mass $m$ sitting on the surface of Earth. Therefore, these two expressions are equal. Set these two formulas equal to each other and solve for Earth's local gravitational field strength $g$ in terms of the other constants.

**c.** The expression from part (b) gives the local gravitational field strength at Earth's surface (a distance $R_E$ from the center of Earth). Plug in numbers for the constants and check to see that you get the expected value of 9.8 N/kg. How would this value change if you were to add 3 m to Earth's radius, which would give the value of $g$ at the ceiling?

**d.** The expression you obtained in part (b) can be generalized to give the local gravitational field strength at an arbitrary distance from the center of Earth. All we need to do is replace $R_E$ by an arbitrary distance $r$. Make this substitution, and then plug in the appropriate values to calculate the gravitational field strength on the International Space Station, which orbits at an average *altitude* of 400 km.

**e.** Now determine *Earth's* gravitational field strength when you are on the moon, which orbits at approximately 384,000 km from the center of Earth.

**f.** Lastly, calculate the *moon's* gravitational field strength on its surface by using the values $M_{\text{Moon}} = 7.35 \times 10^{22}$ kg and $R_{\text{Moon}} = 1,740$ km. (Your value should be significantly larger than Earth's gravitational field strength at this location.)

---

Newton's law of universal gravitation is very general and applies to all masses in the universe. But as previously mentioned, gravitational forces tend to be extremely weak, so it is only when astronomical objects are nearby (stars, planets, moons, etc.) that we need to worry about such forces. Thus, when discussing physics on the surface of Earth, it is much easier to use the simpler formula $F_g = mg$, where $g = 9.8$ N/kg. As one last application of Newton's law of universal gravitation, we return to the topic of orbital motion and derive Kepler's third law.

### 6.10.3. Activity: Kepler's Third Law

**a.** In Activity 6.9.2, we found that the speed needed to throw a ball (horizontally) so that it would end up orbiting Earth (assuming no air drag, no buildings, a perfectly smooth and spherical Earth, etc.) was $v_0 = \sqrt{gR_E}$. This velocity assumes that the ball is thrown just above the surface of Earth (at a distance of $R_E$ from the center of Earth). But we also learned from Newton's law of universal gravitation that $g = GM_E/R_E^2$. Substitute this expression for $g$ into the orbital velocity formula and write it below.

**b.** This new velocity depends on two constants ($G$ and $M_E$) and is technically valid only at the surface of Earth (a distance of $R_E$ from the center of Earth). To make it valid at other distances, we need to change $R_E$ to some other distance. Thus, replacing $R_E$ by an arbitrary distance $r$ leads to a formula that is valid at any distance. Write this new formula below.

**c.** In Activity 6.9.3, we saw that the orbital period $T$ could be calculated by dividing the circumference of the orbit by the velocity. Follow this procedure to determine the orbital period when an object orbits at an arbitrary distance $r$ from the center of Earth.

**d.** Congratulations, you have just derived Kepler's third law (see Eq. (1.1))! If you now square both sides, you will see that the square of the period is proportional to the cube of the orbital distance. Do this below. This is how Kepler's third law is usually stated in words.

# UNIT 7: APPLICATIONS OF NEWTON'S LAWS

*These race cars are traveling around this track at close to 200 miles per hour and are accelerating even when traveling at a constant speed. What forces are responsible for this acceleration, and what keeps the cars from sliding off the track? In this unit we will see how Newton's second law can be used to analyze this situation, along with many others.*

# UNIT 7: APPLICATIONS OF NEWTON'S LAWS

## OBJECTIVES

1. To consider the characteristics of three different types of passive forces: normal, tension, and friction.

2. To learn to use free-body diagrams and Newton's second law to predict the behavior of systems experiencing multiple forces in more than one dimension.

3. To study Newton's third law and its consequences.

4. To explore the phenomenon of uniform circular motion and the accelerations and forces needed to maintain it.

## 7.1  OVERVIEW

In the last unit, we made use of Newton's second law to analyze and understand projectile motion. In this unit, we will apply Newton's second law to several other situations and predict the resulting motion (or the conditions for no motion). The general techniques that we will study are used in many different areas, including astrophysics, engineering, and the study of human body motion.

We begin by considering the characteristics of several common forces that must be included when applying Newton's laws to problems in the real world. These forces—*normal, tension,* and *friction*—are called *passive reaction forces* because they only act in response to other forces. We then make use of free-body diagrams and Newton's second law to learn how to solve several types of problems.

We will also explore the fact that forces are the result of an *interaction* between two objects, and this ultimately leads to Newton's third law. Finally, we will explore *uniform circular motion,* in which an object moves at a constant speed while traveling in a circle. This situation provides a new challenge because the object experiences a continuous acceleration even when moving with a constant speed. We will develop a mathematical description of this *centripetal* acceleration and determine the force(s) necessary to maintain such motion.

## NEWTON'S SECOND LAW AND PASSIVE FORCES

### 7.2  THE NORMAL FORCE

We begin with one of the simplest possible situations: an object sitting at rest on a table. Applying Newton's second law in this situation is straightforward, but we go through the example step-by-step to illustrate a systematic problem-solving procedure that is useful whenever tackling problems using Newton's second law. To perform the experiments in this section, you will need the following equipment:

- 1 wooden block (or other similar object)
- 2 metersticks
- 2 bricks or large books (for raising ends of the meterstick above the table)
- 1 1-kg mass
- 1 piece of soft foam (optional)

---

### 7.2.1.  Activity: Determining the Normal Force

**a.**  Set a wooden block on the table so that it sits motionless. We know that there is a gravitational force acting on the block, but since the block is not accelerating downward, there must be another force acting on the block. What do you think is responsible for this force, and how exactly does it come about? **Hint**: What is physically touching the block?

**b.**  Because the table is the only thing touching the block (besides air, which we will ignore), it is reasonable to assert that the table is exerting a force on the block. In fact, the surface of the table can exert forces both parallel and perpendicular to the surface; for now, we consider only the perpendicular force, which we call the *normal* force (in addition to its regular definition, the word normal can mean perpendicular). The table exerts a force wherever it touches the block—over the entire bottom surface—but for simplicity we draw it as a single force that acts at the center of the bottom surface of the block (see Fig. 7.1). Because we don't know the strength of this force, we represent its (unknown) magnitude by $F_N$. Begin by drawing a free-body diagram (FBD) of the block below. As always, begin with the gravitational force and then look to see what's touching the object (as mentioned, you can ignore any air forces).

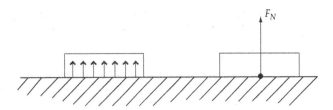

**Fig. 7.1.** For simplification, many small force vectors supporting the bottom of the block are replaced by a single large force vector acting through the center of the block.

   **c.** To apply Newton's second law, we need to choose a coordinate system. Sketch a standard coordinate system to the right of your FBD above, and then apply Newton's second law in the $y$-direction. That is, add up the force components in the $y$-direction and set this equal to $ma_y$ (as always, be mindful of signs).

   **d.** Once we have written down Newton's second law, we can start filling in the information we know. We will assume the mass of the block is known (or can easily be measured), and we know the block is at rest, so the acceleration is zero. Use this information to solve for the normal force magnitude, and then write the normal force using vector notation (components and unit vectors).

We began with this simple problem to illustrate a systematic procedure: Draw a FBD; choose a coordinate system; apply Newton's second law; and finally, make use of the given information to solve the problem. As you gain experience, you may not need to explicitly carry out each of these steps. For example, in the problem above, we know there are only two forces acting and the object remains at rest. Because there's no acceleration, the net force must be zero, which means the normal force must be equal in magnitude and opposite in direction to the gravitational force. Thus, if $\vec{F}_g = -mg\,\hat{y}$, then we can immediately deduce that $\vec{F}_N = +mg\,\hat{y}$. While there's nothing wrong with skipping steps once you've developed sufficient expertise, most errors result from overlooking important information that would be uncovered by following a systematic procedure.

   Let's take a moment to think about the result we just found. You might find it a bit surprising that the normal force has exactly the same magnitude as the gravitational force. How does the table "know" to push back with *exactly* this force? After all, the table can't possibly know the object's weight! Moreover, if you replace the object with one that is heavier (or lighter), the table will respond by pushing back with a larger (or smaller) force such that the normal force is again *exactly* equal in magnitude to the gravitational force. The normal force is an example of a *passive* force, and the following activity explores the mechanism that causes this force.

### 7.2.2. Activity: What Is the Underlying Cause of the Normal Force?

**a.** Use a *single* meterstick as a bridge between two bricks or large books. Gently press down on the center of the meterstick and describe what you observe. What happens to the meterstick? Try pressing with slightly more or less force. Explain what you observe.

**b.** Place a fixed mass on the center of your "bridge" and observe (or measure) how much the meterstick flexes (a 1-kg mass works well for a typical meterstick). Now, add a second meterstick so your bridge consists of two metersticks stacked on top of each other (see Fig. 7.2). With the 1-kg mass at the center, how does the amount of flexing compare to when the bridge consisted of only a single meterstick? Briefly explain what you observe and why you think this is happening.

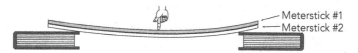

**Fig. 7.2.** Pressing down on two metersticks suspended between textbooks.

**c.** What do you think you would observe if your bridge consisted of three metersticks? What about 10 metersticks? Will there always be at least *some* flexing of the bridge? Explain briefly.

**d.** Do you think there might be something similar going on when a 1-kg mass is placed on the table? Give a brief explanation for how you think the normal force arises when an object is placed on a table (or any other surface). **Hint**: Think about what's happening on a molecular scale; you might try placing the 1-kg mass on a piece of soft foam and observe what happens.

---

As you probably surmised, the normal force is due to the response of the surface on which the object sits. Just as the meterstick bends with weight on it, the table surface also deforms (ever so slightly) in response to an object sitting on it. How much a surface deforms depends on the details of the molecular structure of the material (and is the domain of materials science). But even without knowing the details, we can use this idea to understand how the normal force seems to know exactly how hard to push back on an object.

Imagine placing an object on the table and assume that the normal force is initially zero. What would happen? Well, because the gravitational force pulls down on the object, Newton's second law tells us that the object should accelerate downward. This downward motion will cause the table to flex (a tiny amount) and will result in an upward normal force, just like the flexing of the meterstick results in an upward force. (You can think of the molecules in the table being connected by tiny springs.)

David P. Jackson

If the normal force is smaller than the gravitational force, the object will continue accelerating down into the material, deforming it even more. This additional deformation will result in the material flexing even more, causing an increase in how much the material pushes back. On the other hand, if the normal force happens to be larger than the gravitational force, then Newton's second law tells us that the object will accelerate upward, which tends to decrease the amount of flexing and cause a reduction in how much the material pushes back. The result is that the material responds by flexing until a perfect balance between the two forces is achieved, at which point no further movement will occur.

A brief summary of the normal force:

- A normal force from a surface is *passive*, in that it *reacts* to another force pressing against the surface.
- A normal force acts *perpendicular* to the surface (it pushes the adjoining object *away*). This is true even if the surface is not horizontal or if the surface happens to be curved.
- A normal force can take on different values, depending on how hard an object presses against the surface.
- A normal force for an object on a surface is *not* always equal to *mg*. (Beware, this is a common misconception!)

## 7.3   THE TENSION FORCE

Another common force is a *tension* force. A tension force is typically exerted by a string while pulling on an object, but it can also be exerted by ropes, rubber bands, bungee cords, or even rigid rods. To perform the experiments in this section, you will need the following equipment:

- 2 force sensors
- 1 wooden block with hook or eyebolt on top (or other similar object)

**Fig. 7.3.** Pulling on a box with a rope utilizes a tension force.

- 2 pieces of string (10 cm and 50 cm)
- 1 pulley
- 1 table clamp
- 1 rod
- 1 large rubber band
- 1 platform scale
- 1 spring scale (5 or 10 N)
- 1 1-kg mass

### 7.3.1. Activity: Tension in a String

**a.** Connect two force sensors with a short piece of string (the force sensors will be pulling in opposite directions, so you should *reverse the direction for one of them* so that the pull is negative). After zeroing the sensors in the correct orientation, start the software and gently pull on the sensors with varying forces in a game of tug-of-war. Describe how the force on one end of the string (as read by one force sensor) compares to the force on the other end of the string (as read by the other force sensor).

**b.** Replace the short string with a longer string and repeat the experiment. Do you notice any change in the results?

**c.** Now place the string over a pully so that one force sensor is oriented vertically and the other is oriented horizontally. After *zeroing the sensors in the new orientations*, repeat the experiment and describe what you observe.

There are a few important results from this experiment. First, you should have noticed that a string can only exert a force when under *tension*, or when it is pulled taut. In addition, the force is always directed along the string so that the direction of the string is precisely the direction of the force. You should also have noticed that a string can exert forces of different magnitudes, but it does so in response to some other force (such as the force you are pulling with). Thus, like a normal force, a tension force is another example of a *passive* force. Lastly, it should be clear that the magnitude of the force exerted by the string is the same on both ends of the string, and this is true regardless of the length of the string or whether the string goes over a pulley.

### 7.3.2. Activity: Determining the Tension Force

**a.** Use a piece of string to lift a small object in the air and hold it motionless. As in Activity 7.2.1, draw a FBD of the hanging object. As always, begin with the force of gravity and then look for anything that's touching the object. (It is true that there is air touching the object, but as usual we will ignore the effects of the air.) Draw arrows representing these forces and label them by their magnitudes, using $F_T$ to represent the tension force.

**b.** Choose a coordinate system and sketch it next to your FBD in part (a). Then write down Newton's second law in the vertical direction, keeping all quantities in terms of variables like $m$ and $g$. Finally, use the fact that the object is not accelerating to determine the magnitude of the tension force, and then write down the tension force using vector notation.

This situation should seem similar to the normal force example. Because the object remains at rest (not accelerating), the tension force must exactly balance the gravitational force (just like the normal force in Activity 7.2.1). Although this is a very simple example—and perhaps you even guessed the answer—our main purpose is to utilize our systematic problem-solving procedure as practice for more complicated situations.

In Activity 7.2.2, the visible flexing of the meter stick helped us understand the nature of the normal force from the table. We would also like to understand the underlying cause of the tension force.

### 7.3.3. Activity: The Underlying Cause of the Tension Force

In Activity 7.3.2, you found the tension force had a magnitude of $mg$ when the mass was not accelerating, precisely balancing the gravitational force. Thus, if you're lifting a small mass the tension force will be small, but if you're lifting a larger mass the tension force will be larger. Somehow, the string appears to "know" just how hard to pull to exactly balance the force of gravity. How do you think the string is able to accomplish this task? **Hint**: Consider lifting the object using a rubber band instead of a string, and recall how we explained the underlying cause of the normal force.

Tension forces in materials like a rubber band and string arise due to the stretching response of the material when under tension. While this stretching is easily visible in a rubber band, it is also present in a string (and even in a steel wire) due to the interatomic forces in the material. As with the normal force, you can think of the intermolecular bonds as if they are tiny springs. When you pull on the string, the molecular bonds are stretched (a very tiny amount), and the restoring force increases.

Here is a brief summary of the tension force:

- A tension force in a string has the same magnitude everywhere along the string, even if the string goes over a pulley.[1]
- A tension force from a string is *passive*, in that it *reacts* to another force pulling on the string.
- A tension force acts *along the direction* of the string (it *pulls* on the attached object).
- A tension force can take on different values, depending on how hard one pulls on the string.
- A tension force from a string can only *pull*, it cannot push or exert a lateral (shear) force.[2]

Now that we have some experience with both normal and tension forces, let's try combining them.

---

### 7.3.4. Activity: Combining the Normal Force and the Tension Force

**a.** Consider the following situation (*no experiment necessary yet*): an object of mass $m$ is placed on a table, a string is attached to the top of the object, and someone pulls upward with a constant force $F_T$ that's *smaller* than the gravitational force acting on the object (it remains at rest on the table). Draw a FBD of this situation below, neglecting the air as usual. **Hint**: There are three forces to consider.

**b.** Sketch a coordinate system next to your FDB, and then use Newton's second law in the vertical direction to find the magnitude of the normal force acting on the mass. Your answer should be in terms of symbols.

---

[1] We note this is strictly only true for pulleys that are frictionless and massless. Clearly, any real pulley will be neither, but if the friction is low and the pulley has a low mass in comparison, it is a good approximation.

[2] One can, of course, push on an object with a rigid rod, such as a metal bar. Thus, a rigid rod can exert both a pulling force and a pushing force, and they both act along the direction of the rod. Both forces are due to a material response: when pulling the rod is stretched by a tiny amount, and when pushing the rod is compressed by a tiny amount. (A rigid rod can also exert a shear force, but this is not something we will explore in this class.)

**c.** Now, assume that the object has mass $m = 1$ kg and the upward force is $F_T = 4$ N. Use your formula in part (b) to calculate the value of the normal force in newtons.

**d.** Now try the experiment. Place a 1-kg mass on a platform scale (without attaching the string). **Note**: If your scale reads in mass units (kilograms), you will need to multiply by $g$ to get the force (in newtons). Report the reading on the scale.

**e.** Next, use a force sensor and string to gently pull up on the mass with a force of approximately 4 N. What is the reading on the scale now? How does this compare to your answer in part (c)?

You should have found that the normal force in this situation is *not* equal to the weight of the object ($F_N \neq mg$). In fact, there are many situations in which the normal force is not equal to an object's weight. In this example, the weight of the object is balanced by the normal force *and* the tension force, and by adjusting the upward tension force, the normal force can have any value between zero and $mg$.

## 7.4 FRICTION FORCES

We have seen that when an object is in contact with a surface, the surface can exert a force that's *perpendicular* to the surface. In addition, the surface can also exert a force that's *parallel* to the surface. The force perpendicular to the surface is called the normal force, while the force parallel to the surface is typically called the *friction force* (or just friction). Like the normal force, the friction force is passive in that it is a response to other forces or actions.

While the specific details underlying friction are quite complicated, we will describe a simple model that works surprisingly well. There are two specific cases to consider: *static* friction, when the object remains at rest, and *kinetic* (or *sliding*) friction, when the object slides along a surface. We begin by investigating static friction.

To perform the experiments in this section, you will need the following equipment:

- 1 wooden block with hook or eyebolt on side (or other similar object)
- 1 flat piece of rubber (or other tacky surface)
- 1 spring scale (5 or 10 N)
- 1 platform scale
- 1 500-g mass
- 2 1-kg masses
- 1 data-acquisition system
- 1 force sensor

---

### 7.4.1. Activity: Static Friction

**a.** Place a wooden block on a piece of rubber (or other tacky surface) and attach a spring scale through the eye bolt on the *side* of the wooden block. Gently pull sideways (horizontally) on the spring scale, keeping the force small enough so that the block does *not* move. The force that prevents the object from moving is what we call the static friction force. Try changing the magnitude and direction of your applied force (again making sure that the block doesn't move). Explain how the static friction force seems to respond to the magnitude and direction of your pull.

**b.** Draw a FBD for this situation. Label your applied force (the one from the spring scale) as $F_{app}$ and the static friction force as $f_s$ (friction forces are conventionally written using a lowercase $f$). **Hint**: Even neglecting the air, you should have *four* total forces.

**c.** Sketch in a coordinate system next to your FBD in part (b) and use Newton's second law in the *horizontal* direction to determine the magnitude of the static friction force in terms of the other quantities (assume the block remains at rest).

**d.** If you pull slightly harder or softer, how does the static friction force respond? If you continue to increase the magnitude of your pulling force, what will *eventually* happen? What does this tell you about the static friction force? Explain briefly.

As this activity demonstrates, the static friction force can change both its magnitude and direction, depending on the applied force. While this might seem somewhat surprising, as with the normal force the static friction force is passive and responds to an applied force, and therefore its magnitude (and direction) can change as a result. If you pull gently, the static friction force responds with an equal and opposite force so that the block remains stationary. If you pull harder, the static friction force increases.

Of course, the static friction force cannot get arbitrarily large; if you pull hard enough, the block eventually starts to move! This implies there is a *maximum value* for the static friction force.[3] Once the applied force exceeds the maximum static friction force, the object begins to slide. At this point, the static friction force is replaced by a sliding (or kinetic) friction force.

A brief summary about the static friction force:

- A static friction force is *passive*, in that it reacts to other forces acting on an object.
- A static friction force acts *against* the direction of *impending* motion. In other words, a static friction force acts opposite the direction the object *would* move if there were no friction acting.
- A static friction force can take on different values (and different directions), depending on the strengths (and directions) of the other forces.
- There is a maximum value of the static friction force beyond which the object starts to move.

Once the object starts to move on a surface, static friction stops acting and the object experiences another friction force, $f_k$, called *kinetic friction* (or *sliding friction*). Kinetic friction is relatively easy to investigate. As we will see, the kinetic friction force depends on the interaction between the object and the table (e.g., how hard the object pushes down on the table).

---

### 7.4.2. Activity: Kinetic (or Sliding) Friction

**a.** Using a force sensor and some string, pull a wooden block across the table at a slow, steady velocity. It will be difficult to maintain a perfectly constant velocity, so just do your best (you do not need to use a motion sensor in this experiment, just pull at a slow, constant velocity). Below, draw a FBD of the block (neglecting air resistance). Once again, there will be four forces acting on the block.

---

[3] The same is also true for the normal and tension forces. A material surface can only push back with some maximum amount of force before the breaking point of the material is reached. If you push (or pull) hard enough, the material will eventually give way, and there will be no more normal or tension force. Of course, this is easier to achieve for some materials (e.g., a piece of string) than others (e.g., a table). In general, we will assume that a surface is always able to support the normal force and that a string (or rope) can always support the tension force.

**b.** Sketch a coordinate system next to your FBD and use Newton's second law in the horizontal direction to show that the kinetic friction force is equal to your pulling force (but in the opposite direction).

**c.** We now want to repeat the experiment a number of times while increasing the friction force. We can do this by adding mass to the block, which increases the force pushing down on the table due to the weight of the mass-block combination. Carry out this procedure and fill in the table below.

| Weight of object (N) | Friction force (N) |
|---|---|
| | |
| | |
| | |
| | |
| | |

**d.** If you look at the data in the table above, you might notice that the friction force appears to increase proportionally with the force pushing on the table (the weight). Find the proportionality constant by plotting the friction force as a function of the weight, and then fitting a straight line to the data. Sketch your graph and report the constant of proportionality below.

---

The proportionality constant in this experiment depends on the two surfaces that are interacting. This should not be too surprising given that different materials have different physical properties. We call the proportionality constant the *coefficient of kinetic friction* and represent it using the symbol $\mu_k$ ($\mu$ is the Greek letter mu, pronounced "mew"). In this experiment we plotted the friction force versus the weight of the object. But as we learned in Activity 7.2.1, for a fixed horizontal surface and no other vertical forces acting besides gravity, the normal force has the same magnitude as the object's weight. Because both the normal force and the friction force act on the object, it is conventional to write the kinetic friction force in terms of the normal force. The magnitude of the kinetic friction force is therefore written as

$$f_k = \mu_k F_N \tag{7.1}$$

One of the nice things about kinetic friction is that it is constant for a particular situation. As long as you know the normal force, you can immediately calculate the kinetic friction force (as long as you know the coefficient of friction). The same is not true for static friction. Recall that the static friction force can vary from zero up to some maximum value. In analogy with kinetic friction, we define the *coefficient of static friction* $\mu_s$ in terms of this maximum force:

$$f_s^{\,max} = \mu_s F_N \tag{7.2}$$

The magnitude of the static friction force is therefore always less than or equal to this maximum value; in other words, the static friction force obeys the inequality

$$f_s \leq \mu_s F_N = f_s^{\,max} \tag{7.3}$$

**Note**: Even if you know the normal force and the coefficient of static friction, the static friction force *cannot*, in general, be calculated. It is only the *maximum value* of the static friction force that can be calculated from Eq. (7.2).

In the previous activity, we were able to determine the coefficient of kinetic friction experimentally, but we did not determine the coefficient of static friction. The way the coefficient of static friction is generally calculated is to slowly increase an applied force on an object until it starts to move. The maximum force can then be used to determine the coefficient of static friction using Eq. (7.2) (see Fig. 7.4).

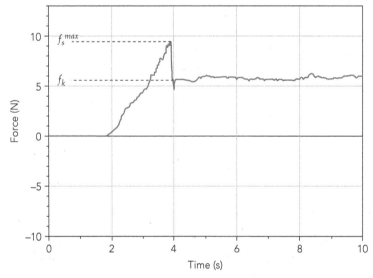

**Fig. 7.4.** A block on a table is pulled using a force sensor. The block remains motionless until approximately 4 seconds, at which point it moves with a (reasonably) constant velocity.

Interestingly, the coefficient of static friction turns out to be greater than the coefficient of kinetic friction:

$$\mu_s > \mu_k$$

Because of this relation, the maximum static frictional force is greater than the kinetic friction force. You may have experienced this when trying to push a heavy object across the floor (or even in the previous experiment); it can be difficult to get the object moving, but once it starts moving you don't have to push quite as hard to keep it moving.

## 7.5 PROBLEM SOLVING: PART 1

In the previous sections we studied three types of common, passive forces. In addition, we have used FBDs to show the different forces acting on objects. You are now ready to use these items, along with Newton's second law, to solve some (relatively) real-world problems.

In each of the problems below, be sure to:

1. Draw a clear FBD.
2. Define your coordinate system.
3. Apply Newton's second law in the horizontal and/or vertical directions.
4. Solve the original problem.

**Note**: Try to use variables instead of numbers whenever possible. Once you have a solution in terms of variables, you can plug in specific values to obtain a numerical solution. Although this approach may seem awkward at first, it is generally much more efficient overall.

### 7.5.1. Activity: Doing Pull-ups

A student with a weight of 132 pounds (mass of 60 kg) is at the fitness center doing pull-ups on a bar: They reach up and grab a bar over their head, hang from the bar by their hands, and then slowly pull themselves up until their chin is at the same height as the bar.

**a.** Approximately how much force (in newtons) must their arms exert to *slowly* lift themselves up to the bar. Assume *slowly* here implies that their velocity on the way up is essentially constant.

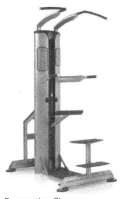

Freemotion Fitness

**b.** After doing a few pull-ups, the student realizes they are too difficult, so they resort to doing "machine-assist pull-ups." They kneel on a platform and select an assist weight, which results in a tension force in a cable that pulls up on the platform. During a pull-up, the cable pulls upward on the platform with a tension force, and the platform pushes upward on the student's body with a normal force (equal in magnitude to the tension force). Although the cable-platform system on the machine may be quite complicated, it is only the platform that is in contact with the student. If the student sets the tension force in the cable to be 50 pounds, how much force must their arms exert to slowly lift themselves up now?

### 7.5.2. Activity: Moving a Box

A box of mass 20 kg is at rest on the floor. The coefficients of friction between the box and the floor are $\mu_s = 0.20$ and $\mu_k = 0.12$. You apply a force to the box (a push) in the horizontal direction to move it across the room.

a. How hard do you need to push to get the box moving? **Hint**: You might find it easiest to think about the *maximum* amount of force you can push on the box *without* it moving. This is essentially the value we're interested in, since even the tiniest additional amount will cause it to move.

b. After the box starts to move, you keep pushing with the *same* force as in part (a). How long, in seconds, will it take you to move the box 4.0 m across the floor. **Hint**: Does the object accelerate at a constant rate? If so, can you make use of the kinematic equations?

## THE TENSION FORCE, ATWOOD'S MACHINE, AND NEWTON'S THIRD LAW

We introduced the tension force as an example of a passive force that only acts in response to another applied force and considered simple cases where the acceleration was zero. In the following sections, we will explore some additional situations involving the tension force, including circumstances that lead to an acceleration.

To perform the experiments in this section, you will need the following equipment:

- 2 large-force spring scales with ropes (or handles) on both ends
- 1 table clamp
- 1 rod

### 7.6   ACTIVE VERSUS PASSIVE PULLING

Suppose you find yourself in the situations depicted in Figs. 7.5a and 7.5b. You are holding onto two ropes, which are attached to spring scales that allow you to read the tension in each rope. In the first situation, two people actively pull outward on ropes attached to spring scales, each of which reads 150 N. In the second situation, one of the ropes is tied to a nearby tree, so the pulling is passive. You remain at rest in both situations.

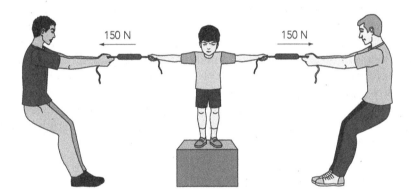

Fig. 7.5a  Being pulled in two directions by two people.

Fig. 7.5b  Being pulled by one person, while the other spring scale is tied to a tree.

### 7.6.1. Activity: Pulling Versus Being Held

**a.** Which situation do you think will cause you the most discomfort, that shown in Fig. 7.5a or 7.5b? In other words, do you think there will be any difference in what you feel in these two situations? Explain the reasons for your prediction. **Hint**: Consider applying Newton's second law to determine the unknown force reading in Fig. 7.5b.

**b.** Working with your group, set up both situations using large spring scales and ropes to test your prediction. What do you find? Can you tell the difference between the two situations?

The previous example is surprising to many people. It is commonly thought that if two people are pulling in opposite directions with a given force, then the "total" force in the string (or on whatever object is between the people who are pulling) will be twice as large. However, if the person in the middle of Fig. 7.5 has their eyes closed, they would have no way of knowing if the rope is connected to a wall or to another person pulling in the opposite direction, as they experience exactly the same thing in both situations. Applying to Newton's second law, we know that the net force acting on the person in the middle in both situations is zero, which means the force pulling to the left and the force pulling to the right must be equal, regardless of whether the force on the right is being supplied by a person or a tree!

## 7.7  TENSION AND ACCELERATION

Now that we have some experience using Newton's second law, it's time to look at a more challenging example. In the following activity we investigate what happens when a cart is attached to a string that runs over a pulley and is connected to a hanging mass. To perform the experiments in this section, you will need the following equipment:

- 1 dynamics cart
- 1 force sensor (attached to cart)
- 1 motion sensor
- 1 hanging mass (approx. 200 g)
- 1 length of string (approx. 1 m)
- 1 cart track with pulley at end (2 m)
- 1 data-acquisition system
  OPTIONAL (Atwood's machine):
  - 1 table clamp
  - 2 rods

- 1 90° clamp
- 1 pulley
- 1 length of string (approx. 1 m)
- 1 set of hanging masses
- 1 data-acquisition system
- 1 motion sensor

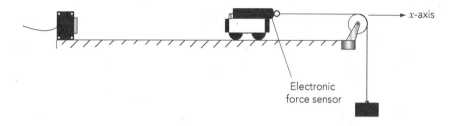

### 7.7.1. Activity: Accelerating Cart

**a.** Consider the situation where a string connects a hanging mass of approximately 200 g over a pulley to a force sensor attached to a (frictionless) cart (see figure above). Starting with everything at rest, we release the cart and use a motion sensor to measure the (horizontal) acceleration of the cart along the track. *Before* setting up the experiment, make a *prediction* about the cart's acceleration.

1. The cart will accelerate to the right at a rate of 9.8 m/s².
2. The cart will accelerate to the right at a rate *less than* 9.8 m/s².
3. The cart will accelerate to the right at a rate *more than* 9.8 m/s².
4. The cart will move to the right with a constant velocity.

**b.** Next, make a *prediction* about the tension in the string. Specifically, imagine the cart is initially held at rest with the mass hanging freely before it is released. Compare the tension in the string when the cart is in motion to the tension in the string when the cart is being held at rest.

1. The tension when the cart is in motion will be *less than* the tension when at rest.
2. The tension when the cart is in motion will be the *same as* the tension when at rest.
3. The tension when the cart is in motion will be *greater than* the tension when at rest.
4. The tension when the cart is in motion will be zero.

**c.** Now set up the experiment. Before attempting to take data, make sure the motion sensor will track the cart over the entire range of motion. Don't forget to *zero the force sensor* when the string is slack. Then, hang the mass over the pulley while holding onto the cart so it doesn't move. After you start collecting data, *continue holding the cart for a couple of seconds before releasing it* so that you measure the tension when the cart is at rest. Then let go, being sure to stop the cart before it crashes into the pulley. You might need to try the experiment a couple of times to get good data. Once you have clean data, make a rough sketch of the force-time graph below.

**d.** You should be able to identify two different regimes to the data: the region before you let go (when everything is stationary) and the region after you let go (when everything is accelerating). For *each* case, use the experimental data to determine (1) the tension in the string and (2) the acceleration of the cart. Write these results below.

**e.** Consider the first region, when everything is stationary. Use what we have learned thus far to explain the results for the tension and acceleration in this region. (This should not be terribly difficult.)

**f.** Now consider the second region, when everything is moving. Is the tension the same as when the cart is stationary? What about the acceleration? How do these results compare to your prediction?

**g.** During the time the cart is moving, compare the tension force to the product of the cart's mass times its acceleration. How do these values compare? (They should be reasonably close, but probably won't be exactly equal.) Can you think of any reason why there might be a small discrepancy between these two values? **Hint:** Might there be another force we are neglecting?

---

Many people are surprised by these results. It is common to think that the tension in the string will be the same when the cart is moving as when it is stationary. Similarly, some people feel the acceleration should be the same as when an object is in free-fall. But notice that the acceleration is significantly smaller than the free-fall acceleration. Fortunately, we can understand exactly what's happening by carefully applying Newton's second law.

### 7.7.2. Activity: Analyzing the Accelerating Cart Experiment

**a.** We will now apply Newton's second law to the above experiment, focusing on the time *after the cart has been released*. Begin by drawing *two* FBDs: one for the cart (of mass $m_1$) and one for the hanging mass (of mass $m_2$). For now, label the tension forces on the different masses as $F_{T_1}$ and $F_{T_2}$ (or $T_1$ and $T_2$ if you prefer). You should ignore any friction acting on the cart. **Hint**: You should have five total forces: three acting on the cart and two acting on the hanging mass. Also, sketch in a standard coordinate system next to your FBDs.

**b.** We are mainly interested in the direction of motion. For the cart, write down Newton's second law in the $x$-direction, while for the hanging mass, write down Newton's second law in the $y$-direction. Note that we need to use the $x$-component of the cart's acceleration $a_{1x}$ (for $m_1$) and the $y$-component of the hanging mass's acceleration $a_{2y}$ (for $m_2$). Just focus on writing down the equations in terms of the variables; you shouldn't put in any numbers or solve anything yet!

**c.** Based on our earlier activities, what can you say about the *magnitudes* of the tension forces, $F_{T_1}$ and $F_{T_2}$, in this experiment. Explain briefly.

**d.** Based on what you observed while the experiment took place, what can you say about the *magnitude* of the cart's acceleration (in the horizontal direction) compared to the *magnitude* of the hanging mass's acceleration (in the vertical direction)? **Hint**: Does the string change its length while the masses are moving? This is a little bit subtle, so make sure you think it through carefully.

**e.** Now comes the sneaky part. You should have found the magnitudes of the two tensions, as well as the magnitudes of the two accelerations, are the same: $F_{T_1} = F_{T_2}$ and $|a_{1x}| = |a_{2y}|$ (we need to use absolute values here because $a_{1x}$ and $a_{2y}$ are acceleration *components*, which can be positive or negative, and all we know is that their *magnitudes* are equal). We want to use this information in Newton's second law, but we first need to understand the *signs* of the acceleration components. Based on our coordinate system, if the cart is accelerating in the positive $x$-direction, in what direction will the hanging mass be accelerating? What does this tell you about how the *components* $a_{1x}$ and $a_{2y}$ are related mathematically? Briefly explain.

**f.** We now want to rewrite the two equations from part (b) while making use of parts (c), (d), and (e). In particular, rewrite these equations making the following two substitutions based on what we just argued: (i) $F_{T_2} = F_{T_1}$ (because the tension is the same everywhere in the string, we can write both tensions simply as $F_T$), and (ii) $a_{2y} = -a_{1x}$ (replace $a_{2y}$ with $-a_{1x}$).

**g.** You should now have two equations that contain only the variables $m_1$, $m_2$, $g$, $F_T$, and $a_{1x}$, three of which we already know (or can easily measure): $m_1$, $m_2$, and $g$. We therefore want to solve for the two unknowns: $a_{1x}$ (the acceleration) and $F_T$ (the tension). The easiest way to do this is probably by *substitution*: solve one equation for $F_T$, plug this result into the other equation, and then solve for $a_{1x}$. Carry out these calculations below and show that you get $a_{1x} = \left( \dfrac{m_2}{m_1 + m_2} \right) g$.

**h.** To find the tension, substitute your result for $a_{1x}$ back into the first equation and then solve for the tension. Show that you get $F_T = \left( \dfrac{m_1 m_2}{m_1 + m_2} \right) g$.

**i.** Finally, plug in the known (measured) values for $m_1$, $m_2$, and $g$ to calculate numerical values for the acceleration and tension. (If you haven't already measured the cart's mass, you will need to do so.) How do these values compare to what was measured experimentally? Are they reasonably close?

**j.** It can be informative to consider the "limiting cases" for our results. In particular, use the equations for $a_{1x}$ and $F_T$ to determine the acceleration and the tension for each of the situations below. In each case, replace the specific variable with the value indicated and determine the new form of the equation (in terms of variables). *Briefly explain* why the answers makes sense.

    **(i)** $m_2 \to 0$ (essentially removing the hanging mass)

    **(ii)** $m_1 \to 0$ (essentially removing the cart)

    **(iii)** $m_1 \to \infty$ (making the cart "infinitely heavy")

---

The previous activity is fairly detailed and contains a lot of information. If you had some trouble with this example, we recommend you go over it again to make sure you understand all the steps. And if you still have questions, be sure to ask your instructor! The following activity asks you to analyze a similar situation using the same ideas (this new situation is known as Atwood's machine). Instead of guiding you through it step-by-step, we'll let you tackle it more-or-less on your own.

### 7.7.3. Activity: Atwood's Machine

**a.** Atwood's machine consists of two masses, $m_1$ and $m_2 > m_1$, connected by a string that's hanging over a pulley, as shown in the diagram. **OPTIONAL EXPERIMENT**: Working together with your group, set up the experiment and try it out with different mass combinations (not taking any data yet). After a few tries, choose appropriate masses and use the motion sensor to measure the acceleration of *one* of the masses. This is probably best done by placing the motion sensor on the floor "looking up" to measure the position of the rising mass ($m_1$). Write down your measured values for $m_1$, $m_2$, and the acceleration below.

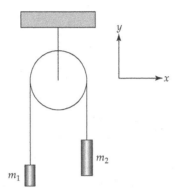

**b.** Perform a theoretical analysis. Begin by drawing a FBD for each mass and then apply Newton's second law in the vertical direction to each mass using the coordinate system in the figure above.

**c.** Next, combine your Newton's second law equations from part (b) to solve for the acceleration of the system. As in Activity 7.7.2, think carefully about how the tension and acceleration components of the two masses are related to each other. Make the appropriate substitutions and then solve for the acceleration in terms of $m_1$, $m_2$, and $g$.

**d.** If you performed the experiment, substitute your values of $m_1$ and $m_2$ into your equation (don't forget to include the masses of any "mass pans"), and compare your calculated and measured accelerations. If they disagree, can you think of a possible explanation for why? (If you didn't perform the experiment, you can skip this part.)

**e.** Once again consider the "limiting cases" for the acceleration. In particular, use your result from part (c) (with the *variables* as opposed to the numbers) to determine the acceleration for each of the situations below. In each case, replace the specific variable with the value indicated and determine the new form of the equation. *Briefly explain* why the answers makes sense.

   **(i)** $m_1 \to 0$

   **(ii)** $m_2 \to 0$

   **(iii)** $m_1 = m_2$

## 7.8 NEWTON'S THIRD LAW

We have seen a number of different *reaction* forces that respond to another, applied force. For example, when you push on the wall, it "pushes back" on you. Similarly, when you pull on a rope attached to another object, the rope pulls back on your hand. If you drop a ball so that it hits the ground, the ground will exert a normal force on the ball during the collision. We are now ready to formalize these ideas to arrive at Newton's third law. The following equipment will be helpful in making some observations:

- 2 electronic force sensors (or 2 spring scales)
- 1 data-acquisition system (if using electronic force sensors)
- 1 table clamp
- 1 rod

- Various options for dynamic pull (using 2 large-force spring scales):
  - 1 kinesthetic cart (to sit on)
  - 1 skateboard
  - 1 pair roller skates
  - 1 cart with cart track

### 7.8.1. Activity: Forces of Interaction

**a.** Set up a static pull situation (nothing moving) using two electronic force sensors as shown in Fig. 7.6 (don't forget to zero the force sensors). Try varying the force that's being applied. Describe what you observe. How do the magnitudes of the two force measurements compare?

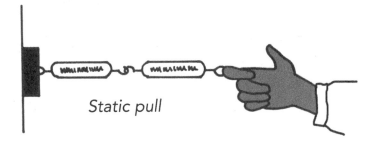

*Static pull*

*Dynamic pull*

Fig. 7.6. Ways to test Newton's third law.

**b.** Finally, try setting up a dynamic situation in which one person (or object) is free to move while being pulled. This may be a bit challenging to do with an actual person on a skateboard, but you can mimic this situation by attaching a force sensor to a cart with some added mass. Again, what do you observe?

**c.** Were there any situations in which the forces did not appear to be (essentially) equal in magnitude? Do you think this will always be true? Explain briefly.

---

This previous activity demonstrates something very simple and yet quite profound. It suggests that any interaction between two objects is accompanied by two related *interaction forces*. Moreover, our observations suggest that these interaction forces have the same magnitude! These ideas are summed up in Newton's third law.

**Newton's Third Law**

Newton's third law can be stated as follows:

> If one object exerts a force on a second object (either through contact or at a distance), then the second object exerts a force back on the first object that is equal in magnitude and opposite in direction to the force from the first object.

Newton's third law can be stated very concisely using vector notation. We can write that the force of interaction of object 2 on object 1 is related to the force of interaction of object 1 on object 2 as:

$$\vec{F}_{2\to1} = -\vec{F}_{1\to2} \quad \text{(Newton's third law)} \tag{7.4}$$

where $\vec{F}_{2\to1}$ represents the force of object 2 on object 1 and $\vec{F}_{1\to2}$ represents the force of object 1 on object 2 (Fig. 7.7).

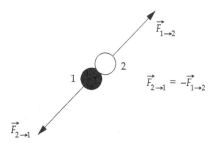

Fig. 7.7. "Equal and opposite" force vectors signifying forces that are equal in magnitude and opposite in direction.

Newton formulated the third law by studying the interactions between objects when they collide. It is difficult to fully understand the significance of this law without first studying collisions, and so we will consider this law again when we discuss collisions.

**7.9** PROBLEM SOLVING: PART 2

### 7.9.1. Activity: Sledding down a Hill

**a.** A student is sledding on an icy hill. As shown in the figure, the hill makes an angle of $\theta = 30°$ with respect to the horizontal, and it is 50 m *along the hill* to the bottom. The student and sled have a combined mass of 63 kg, and the coefficient of kinetic friction between the sled and the ice is 0.1. Begin by adding all the necessary force vectors to the figure to create a FBD.

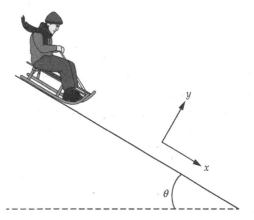

Note that we have chosen a non-standard (or "rotated") coordinate system in the diagram, aligning the $x$-axis along the slope of hill and the $y$-axis perpendicular to the hill. Although any coordinate system is valid, choosing one of the coordinate axes along the direction of acceleration will simplify the calculations (this is worth remembering). In this coordinate system the normal force will point along the $+y$-axis, while the kinetic friction force will point along the $-x$-axis. This leaves the gravitational force vector, which points straight down and does not align with either axis. We therefore need to resolve the gravitational force into its $x$ and $y$ components; in other words, we need to find $F_{gx}$ and $F_{gy}$.

**b.** Below is a figure that shows the gravitational force acting on the sled, along with a redrawn coordinate system that has its origin at the location of the sled. Based on the direction of $\vec{F}_g$ in this coordinate system, it should be clear that $F_{gx}$ is positive and $F_{gy}$ is negative. Using geometry, we can deduce that the angle between $\vec{F}_g$ and the negative $y$-axis is the same angle that the hill makes with the horizontal (if you imagine letting the hill angle $\theta$ get very small, it should be clear that the angle between $\vec{F}_g$ and the negative $y$-axis gets correspondingly small). Using this diagram, determine the components $F_{gx}$ and $F_{gy}$ for the gravitational force (leave in terms of the problem variables).

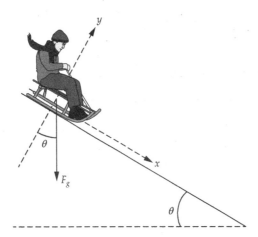

**c.** Because there are forces in both the $x$ and $y$ directions, we need to use Newton's second law in both directions. Let's start by applying Newton's second law in the $y$-direction, leaving everything in terms of variables for now. As usual, on the left side of the equation you should sum up the force components in the $y$-direction to find $F_y^{net}$, while on the right side you should have $ma_y$.

**d.** Your equation in part (c) should contain two unknowns: the normal force and the acceleration in the $y$-direction (we are given $m$, $g$, and $\theta$, and we found $F_{gy}$ in part (b)). Fortunately, our choice of coordinate system makes it easy to determine the acceleration in the $y$-direction (does the sled's $y$-position change as it slides down the hill?). Substitute this acceleration into your equation from part (c) to determine the normal force in terms of the problem variables.

**e.** Now apply Newton's second law in the $x$-direction to determine the acceleration of the sled down the hill. That is, sum up the force components in the $x$-direction, set it equal to $ma_x$, and solve for the acceleration. **Hint**: Remember that the kinetic friction force depends on the normal force.

**f.** Congratulations! By applying Newton's second law in both the $x$ and $y$ directions, we have solved for the acceleration of the sled down the hill. Notice that this acceleration is *constant*, which means we can use the kinematic equations from Unit 4 to determine other aspects of the motion. For example, assume the student started at rest at the top of the hill. How long (in seconds) does it take them to reach the bottom of the hill (go ahead and plug in numbers now)?

**g.** How fast is the student moving when they get to the bottom of the hill?

### 7.9.2. Activity: Treating Connected Objects as a System

A cart of mass $m_2 = 850$ g is initially moving to the right on a track with a speed of $v_0 = 1.8$ m/s. The cart is attached to two masses via strings and pulleys as shown in the diagram. The mass on the left is $m_1 = 250$ g, while the mass on the right is $m_3 = 200$ g. Assume friction is negligible.

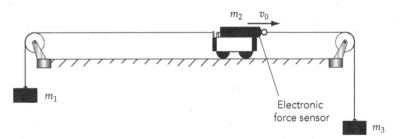

At first glance this seems like a difficult problem. But notice that the set-up is similar to the examples in Section 7.7, so we might be tempted to proceed as we did previously: Apply Newton's second law to each object separately and then combine the results. But as you can imagine, this procedure gets more complicated as the number of objects increases, and it turns out there is an easier way to proceed.

Based on what we learned in Section 7.7, we know that all three masses, being connected by strings, will move together in unison (if they didn't, the strings would either break or bunch up). In some sense, we can think of these three objects as being part of a single object, or a single *system*, composed of the three masses and the strings. Viewed in this way, the strings are *internal* to the system, and it turns out that we don't need to account for any forces that are internal to the system.[4] This idea is no different from ignoring the forces inside a dropped ball: The internal forces act to keep the system components (all the microscopic pieces of the ball) moving together, but have no bearing on the overall motion of the system as a whole.

To proceed, we take the system to be $m_1$, $m_2$, and $m_3$, along with the strings that connect them (assumed to be massless). We start by defining coordinates for the system ("system coordinates"), and we can reduce the complexity of the problem by using what we know about the resulting motion. In this example, as the cart initially moves along the track to the right, $m_3$ will be moving downward, while $m_1$ will be moving upward. We define the positive direction of motion for our system to correspond to $m_1$ moving up, $m_2$ moving to the right along the track, and $m_3$ moving down. (As usual this choice is arbitrary.) In some sense, we are defining an axis that follows the strings connecting the masses, wrapping around from $m_1$ to $m_3$. The simplification occurs because the motion of the system now occurs along a *single* axis (which we typically call the "$x$-axis").

The coordinate system for the system now consists of a "wrapped" $x$-axis, which is aligned with the strings, and a corresponding $y$-axis, which, as usual, is perpendicular to (and counter-clockwise from) the $x$-axis.

**a.** In the diagram above, draw a separate (appropriately oriented) coordinate system next to each of the objects. In other words, put an $xy$ coordinate system next to each object with proper orientation to conform to the description in the preceding paragraph.

**b.** The next step is to identify and label all the forces (essentially drawing three FBDs—one for each object). Begin by sketching and labeling *all* the forces acting on $m_1$, $m_2$, and $m_3$ in the diagram above. **Hint:** There are *eight* total forces.

**c.** In the system approach, *forces internal to the system can be ignored* because they ultimately cancel due to Newton's third law. Internal forces are those that are due entirely to interactions between parts of the system. Of your eight forces from part (b), which ones are internal to the system? (You should notice that these internal forces come in pairs that have equal magnitudes and opposite directions.)

Hopefully, you identified the four tension forces as being internal to the system. The strings, and the tension forces in them, simply serve to connect $m_1$ to $m_2$ and $m_2$ to $m_3$, all of which are part of the system.

---

[4] Technically, we are relying on Newton's third law here. The internal forces come in pairs, and by Newton's third law they all cancel out. We will discuss this in more detail when we consider momentum conservation.

On the other hand, the three gravitational forces are *external* since Earth is not part of the system. Similarly, the track is not part of our system, so the normal force acting on $m_2$ is also an external force. When using Newton's second law to analyze our system, we only need to consider the *external* forces.

In the system approach, we can write Newton's second law as

$$\vec{F}_{\text{ext}}^{\text{net}} = m_{\text{sys}}\vec{a}_{\text{sys}} \quad \text{(Newton's second law in system approach)} \quad (7.5)$$

where $\vec{F}_{\text{ext}}^{\text{net}}$ is the (vector) sum of the *external* forces acting on the system, $m_{\text{sys}}$ is the total system mass, and $\vec{a}_{\text{sys}}$ is the system acceleration. As always, this vector equation is equivalent to (in this case) two component equations. Since we're ultimately trying to find the acceleration, we begin with the $x$-direction.

d. Start by summing the (external) force components in the $x$-direction of the system (there are three). As always, these components can be positive or negative, so we need to refer to the directions of the forces in relation to our "wrapped" system coordinates to determine the signs. As usual, leave everything in terms of variables for now.

e. Now set the sum of these forces equal to $m_{\text{sys}}a_{\text{sys},x}$ and solve for the system acceleration in terms of variables. (The system mass is the total mass of all system components.)

f. Finally, plug in numbers to get a numerical result for the $x$-component of the system acceleration. Does the sign of your answer make sense? Explain briefly.

g. Use your result for the system acceleration to determine how far (in meters) the cart travels before reaching the turning point (the point at which the cart turns around).

## CIRCULAR MOTION AND CENTRIPETAL ACCELERATION

When using Newton's second law to solve problems, we are often in a situation where we know the forces and want to determine the acceleration. However, there have been times when we knew the acceleration was zero, and we used this fact to deduce something about the forces that were acting. In the following sections, we discuss a particular kind of motion that occurs quite frequently: an object moving in a circular trajectory with a constant speed, something referred to as *uniform circular motion*. Examples include a race car speeding around a circular track or a planet like Earth orbiting the sun. As we will see, an object moving in uniform circular motion is continually accelerating, even though its speed is constant.

### 7.10    MOVING IN A CIRCLE AT A CONSTANT SPEED

Suppose you attach a ball to the end of a string and twirl it around so that it moves around in a circle at a constant speed.[5] While quite simple, this motion is clearly not one-dimensional, at least not in the normal sense. Thus, it is important to remember that both velocity and acceleration are vectors.

The following equipment will be needed for the demonstrations in this section:

- 1 ball on string
- 1 kinesthetic cart
- 1 large-force spring scale
- 3 segments of rope

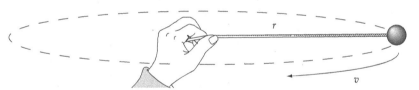

**Fig. 7.8.** Uniform circular motion. A ball moving at a constant speed in a circle of radius *r*.

---

**7.10.1.  Activity: Acceleration in Uniform Circular Motion**

**a.** Consider Fig. 7.8. Determine the speed (in m/s) of a ball that moves in a circle of radius $r = 0.5$ m, assuming it takes 0.30 s to complete one revolution.

---

[5] We will assume the ball is twirled quickly enough that the string is essentially horizontal. Alternatively, you could imagine an astronaut performing this experiment in deep space.

**b.** We have been told that the *speed* of the ball is constant. Does this mean the ball is not accelerating? Why or why not?

**c.** Write down the *definition* of acceleration, taking into account the fact that acceleration is a vector.

**d.** In the situation shown in Fig. 7.8, is the *velocity* of the ball constant? Explain briefly.

**e.** Considering your answers to parts (c) and (d), explain how an object with a constant *speed* can still have an acceleration.

---

Hopefully it's clear that the *velocity* of a ball in uniform circular motion is changing, even though its *speed* is constant. In particular, the *direction of the velocity vector* is continually changing as the ball moves around in circular motion. Because acceleration is defined as the rate of change of the velocity vector, an object will undergo an acceleration whenever its speed or direction changes. In the next activity, we investigate the forces leading to this acceleration.

---

### 7.10.2. Activity: Forces in Uniform Circular Motion

**a.** Consider again Fig. 7.8. For simplicity, assume the experiment is performed by an astronaut in space so that the motion is completely horizontal and gravity and air resistance can be neglected. In the space below, draw *three different* FBDs representing the ball at three different locations in its trajectory. Draw your FBDs *as viewed from above*, so that the ball's trajectory forms a circle. **Hint**: What is the only thing touching the ball?

**b.** Based on your diagrams, in what direction is the net force always pointing? What does this tell you about the ball's acceleration? Explain briefly.

---

Although the previous situation is quite simple, it demonstrates the power of Newton's second law. By drawing FBDs at several times, we see that there is only a single force acting on the ball (tension in the string), and that this force *always* points to the center of the circle. Since the net force points to the center of the circle, we can immediately conclude that the acceleration must also point to the center of the circle. Such an acceleration is often referred to as a *centripetal acceleration* (in this context, the word *centripetal* means *center seeking*).[6]

In the next section, we derive how the centripetal acceleration relates to the speed and radius of the object's orbit. But first, let's experimentally demonstrate that the net force acting on an object in uniform circular motion really does point toward the center of the circle.

### 7.10.3. Demonstration: Forces in Uniform Circular Motion

We will perform a demonstration experiment where a person sits on a wheeled cart (free to move in any direction) and holds a rope that's attached to a fixed, center post (see Fig. 7.9). A spring scale is attached between the center post and the rope, and the end of the rope should be held by the rider. The cart is then pulled by someone else using a second rope so that it moves in uniform circular motion. The second rope should be attached to the cart in a direction perpendicular to the original rope, and the person pulling should do their best to keep the rider moving at a constant speed. This pulling force is needed to overcome significant friction in this experiment. It should be clear to the rider that the central rope exerts a tension force on them that is always directed toward the center of the circle!

---

[6] Similarly, because the tension points to the center of the circle, it is sometimes referred to as a *centripetal force*. However, the term centripetal force can cause a lot of confusion and is unnecessary, so we have chosen to avoid this term altogether.

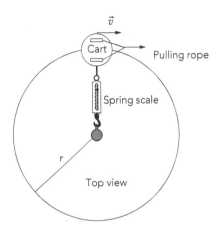

**Fig. 7.9.** Top view of the uniform circular motion experiment.

This experiment can be made quantitative by measuring the distance from the center post to the center of the rider. Then, as the rider is pulled around in a circle at a constant speed, they can read the force from the spring scale while someone else measures the time it takes to travel once around the circle (to get the speed).[7]

## 7.11   THE CENTRIPETAL ACCELERATION

We have seen that an object moving in uniform circular motion is continually changing its *direction* and is therefore accelerating. In the next activity, we will confirm the direction of the acceleration and quantify its magnitude in terms of the speed of the object and the radius of the circle on which it moves. We start by using vector addition to confirm the direction of the acceleration as an object moves around a circle.

### 7.11.1.  Activity: The Direction of Centripetal Acceleration

**a.** Consider the direction of motion of a ball moving clockwise in a circle at a constant speed. The diagram below shows two points (A and B), along with points *just before* and *just after* point A. On this diagram, draw an arrow representing the ball's velocity at the dot *just before* point A. Label this vector $\vec{v}_1$. **Note**: Since the trajectory of the ball is a circle, the ball's velocity is always *tangent* to the circle as it moves.

---

[7] It can be challenging to obtain reliable quantitative data in this experiment, so we will experimentally test the centripetal acceleration formula a different way in Section 7.12.

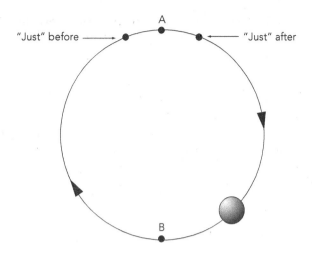

**b.** On the same diagram, draw an arrow representing the velocity of the ball at the dot *just after* it passes point A. Label this vector $\vec{v}_2$. Explain how the length of this vector is related to the length of vector $\vec{v}_1$.

**c.** Now find the vector representing the *change in velocity* $\Delta\vec{v} = \vec{v}_2 - \vec{v}_1$ from *just before* to *just after* point A. One way to do this is to add the two vectors $\vec{v}_2$ and $-\vec{v}_1$ (the vector opposite to $\vec{v}_1$). Another way, perhaps slightly easier, is to recognize that $\vec{v}_2 = \vec{v}_1 + \Delta\vec{v}$, so if you draw the vectors $\vec{v}_1$ and $\vec{v}_2$ with their *tails* together, then the vector $\Delta\vec{v}$ is the vector that goes from the tip of $\vec{v}_1$ to the tip of $\vec{v}_2$. Whichever technique you choose, show how you find the vector $\Delta\vec{v}$. (Try to be fairly accurate here.)

**d.** Now, draw a copy of $\Delta\vec{v}$ on the diagram above, placing the tail of this vector exactly at point A (halfway between the two vectors used to determine $\Delta\vec{v}$). Make sure this vector has the same direction as the vector you found in part (c).

**e.** Based on the definition of acceleration $\left(\vec{a} = \frac{\Delta\vec{v}}{\Delta t}\right)$ and the vector $\Delta\vec{v}$ you just drew, what is the *direction of the acceleration* at point A in the diagram above? Is this vector analysis consistent with the net force pointing to the center of the circle? Explain briefly.

**f.** Imagine you redid the analysis centered about point B at the opposite side of the circle. What do you think you would find for the direction of the acceleration at this point? Explain briefly.

This activity demonstrates how vector addition leads to an acceleration for uniform circular motion that is directed toward the center of the circle. This result is consistent with what we found in the previous activity, namely, that the force acting on an object in uniform circular motion always points toward the center of the circle.

### 7.11.2. Activity: The Magnitude of the Centripetal Acceleration

**a. Step 1: Angles.** Refer to the diagram below. Explain why, at the two points shown on the circle, the angle $\theta$ between the position vectors $\vec{r}_1$ (at time $t_1$) and $\vec{r}_2$ (at time $t_2$) is the same as the angle $\theta'$ between the velocity vectors $\vec{v}_1$ (at time $t_1$) and $\vec{v}_2$ (at time $t_2$). **Hint:** In circular motion, velocity vectors are always tangent to the circle and therefore perpendicular to their position vectors.

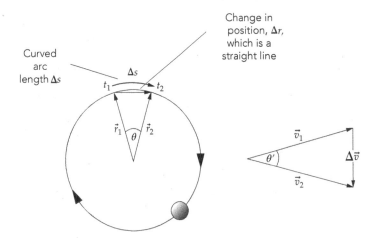

**b. Step 2: Arc Length.** Consider the roughly triangular region defined by the position vectors $\vec{r}_1$ and $\vec{r}_2$ and the arc length $\Delta s$ along the outside of the circle. For an object traveling in circular motion, the vectors $\vec{r}_1$ and $\vec{r}_2$ will have the same magnitude. Use this fact to relate the arc length

$\Delta s$ to the angle $\theta$ and radius $r$, where $r = |\vec{r}_1| = |\vec{r}_2|$. **Hint**: Remember how radians are defined.

c. Similar to part (b), we can consider the triangular region defined by the velocity vectors $\vec{v}_1$, $\vec{v}_2$, and $\Delta\vec{v}$. For an object traveling in circular motion, the vectors $\vec{v}_1$ and $\vec{v}_2$ will have the same magnitude $v = |\vec{v}_1| = |\vec{v}_2|$. Use this fact to relate $|\Delta\vec{v}|$ to the angle $\theta'$ (which is equal to $\theta$) and $v$. **Hint**: You need to assume that the length $|\Delta\vec{v}|$ is the same as the circular arc length that joins the tips of $\vec{v}_1$ and $\vec{v}_2$. (Technically, this is only true if the angle $\theta' = \theta$ is small.)

d. **Step 3: Speed and Distance**. Write $\Delta s$ in terms of the speed $v$ and change in time $\Delta t = t_2 - t_1$. **Hint**: Remember that speed is nothing more than the distance traveled in a given amount of time.

e. You should now have three equations from the previous three parts:

$$\Delta s = r\theta$$

$$|\Delta\vec{v}| = v\theta' = v\theta$$

$$\Delta s = v\Delta t$$

In addition, the magnitude of the acceleration can be written as

$$a = \frac{|\Delta\vec{v}|}{\Delta t}$$

Using the relations above, derive an expression for the magnitude of the acceleration $a$ in terms of only the speed $v$ and the radius $r$. Show your work.

You should have found the following relationship for the magnitude of the acceleration during uniform circular motion (where the subscript "$c$" denotes this is the *centripetal* acceleration):

$$a_c = \frac{v^2}{r} \qquad \text{(Centripetal acceleration magnitude)} \qquad (7.6)$$

This useful formula is valid whenever an object travels in circular motion. If we know the speed and radius of the circular trajectory, we can immediately conclude the object *must* be accelerating toward the center of the circle with a magnitude given by Eq. (7.6). This acceleration is solely due to the change in direction of the velocity vector as the object moves along its circular trajectory.

If an object moves in *uniform* circular motion (with a constant speed), then the centripetal acceleration will be the *only* acceleration experienced by the object. However, in addition to the centripetal acceleration, there can also be a *tangential* acceleration that results if the object changes speed. Specifically, if an object's speed is changing as it moves around the circle, there will be a tangential acceleration component $a_t = \Delta v / \Delta t$ directed tangent to the trajectory (note that $\Delta v$ here represents the change in the object's *speed*, not velocity). Thus, if an object is speeding up, the tangential acceleration is in the same direction as the motion, whereas if it's slowing down the tangential acceleration is opposite the direction of motion.

## 7.12   MEASURING THE CENTRIPETAL ACCELERATION

We just discovered that when an object is moving in uniform circular motion, the centripetal acceleration will be proportional to the square of the velocity and inversely proportional to the radius of the circle. In this section, we would like to test this formula experimentally.

As shown in the diagram, begin by passing some fishing line through a small glass tube about six inches long and 1/8 inch in diameter. A rubber stopper should be tied to one end of the fishing line, and the other end should be tied into a small loop with a mass (75–150 g) hung from the loop. By swinging the rubber stopper around in a circle above your head, you should be able to suspend the weight so that it hangs motionless below the glass tube.

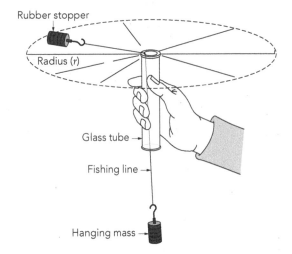

For this measurement, you will need the following equipment:

- 1 rubber stopper
- 1 hanging mass (75–150 g)
- 1 length fishing line (or string)
- 1 glass tube (1/8 inch diameter, approx. 6 inches long)

### 7.12.1. Activity: Centripetal Acceleration Analysis

**a.** Draw a FBD of the hanging mass $m_h$. Assuming this mass is at rest, apply Newton's second law to determine the tension $F_T$ in the string in terms of the problem variables.

**b.** Now draw a FBD (viewed from above) of the rubber stopper (mass $m_s$) at a specific instant of time and choose a coordinate system that has the $x$-axis aligned with the string. (For simplicity, we will assume the string is completely horizontal, but this assumption will only be reasonable if the rubber stopper is being swung around fairly rapidly.) Apply Newton's second law to relate the tension $F_T$ in the string to the $x$-component of the acceleration $a_x$, being careful with signs. (At this point, we are assuming we don't know the acceleration of the stopper, so write it simply as $a_x$.)

**c.** Now, if the rubber stopper moves in uniform circular motion, we know that it must have a centripetal acceleration that points toward the center of the circle with magnitude $v^2/r$. Use this information to replace $a_x$ in your result from part (b) (again, be mindful with signs). Next, use your result from part (a) to substitute in for $F_T$, and then divide both sides of the equation by $m_s$. You should end up with an expression for $v^2/r$ in terms of $m_h$, $m_s$, and $g$.

At this point, you have an equation in which everything can be determined experimentally: One side involves the two masses and the constant $g$, while the other side is the quantity $v^2/r$ (note that both sides of the equation have units of acceleration, or m/s$^2$). The next step is to carry out an experiment to test this equality.

### 7.12.2. Activity: Centripetal Acceleration Experiment

**a.** Before taking any data, practice the experiment a few times. As noted, you want to swing the rubber stopper around in a circle above your head such that the hanging mass does not move up or down during the experiment. There are two important things to remember: (1) You should swing the mass fairly rapidly so that the rotating string remains nearly horizontal (this was one of the assumptions in our analysis), and (2) you want to maintain a steady rhythm whereby the hanging mass remains essentially motionless during the experiment. This will take a little practice.

**b.** Once you feel reasonably confident, work as a team to collect data. The masses are easy to measure when the experiment is finished; the difficult part is measuring the speed and the radius. One person should be in charge of timing how long it takes to make 20 revolutions; this is usually easier if the person swinging the mass counts out loud (perhaps with a bit of a countdown). Someone else needs to be responsible for measuring the radius of the motion. The best way to do this is to measure how far the hanging mass sits below the glass tube while the experiment takes place. Then, when the experiment is over, you can figure out the distance from the top of the glass tube to the center of the rubber stopper.

Carry out the experiment and report the values for the hanging mass $m_h$, the mass of the rubber stopper $m_s$, the radius of the stopper's circular trajectory $r$, and the time it took to complete 20 revolutions. Then calculate the speed of the rubber stopper $v$ by dividing the total distance it traveled by the time you measured.

**c.** Finally, calculate and compare the quantities $m_h g/m_s$ and $v^2/r$. Are they approximately equal? If not, what do you think is the most likely source of the discrepancy: measurement errors, or some approximation we made in the theoretical analysis? If the discrepancy is due to an approximation in the theoretical analysis, which of the two quantities you just calculated should be larger?

Knowing that an object moving in uniform circular motion has a centripetal acceleration given by $v^2/r$ can be helpful when applying Newton's second law. The following problem gives you some practice with this concept.

### 7.12.3. Activity: Using Centripetal Acceleration

**a.** Suppose you perform the previous experiment using fishing line that's rated at 100 Newtons; That is, the line will break if the tension exceeds 100 N of force. Determine the maximum speed that you can swing a 100-g object around in a circle if the length of the string is 1 m.

**b.** Convert your speed in part (a) to rpm (revolutions per minute).

**c.** How would your answer to part (a) change if the mass was doubled? Support your response using equations.

**d.** How would your answer to part (a) change if the radius of the circle was changed to a half meter? Again, support your response using equations.

## 7.13  PROBLEM SOLVING: PART 3

We are now ready to bring together everything we have learned and apply Newton's second law to some more challenging situations. As is often the case when tackling real-world problems, we'll want to make some simplifying assumptions in how we model the situation. Although we provide some guidance and hints in the activities below, they will likely require discussion with your group members to successfully complete. But please ask if questions arise!

### 7.13.1.  Activity: Driving in Circles

A car with a mass 1,200 kg drives on a flat circular racetrack with a radius of 250 m. The engine provides an applied force in the forward (tangential) direction that is exactly balanced by drag forces (air resistance and rolling friction) so that the car keeps moving at a *constant* speed $v$ the entire time. The coefficient of kinetic friction between the tires and the road is $\mu_k = 0.6$, and the coefficient of static friction is $\mu_s = 0.8$.

**a.** Draw a FBD for the car, including an appropriate coordinate system. Draw your diagram showing the car as viewed from the rear (you don't need to worry about drawing the applied force or the drag force/rolling friction, as these forces are perfectly balanced).[8] While two of the forces are straightforward, you will need to think carefully about what force keeps the car moving in a circle. **Hint**: What forces can a surface exert on an object, and how might that come into play here?

**b.** Apply Newton's second law in the vertical direction to determine the normal force acting on the car. (This should be straightforward.)

**c.** Now apply Newton's second law in the horizontal direction. Use this equation to determine the *maximum* speed this car can travel *without sliding off the track*. Give your answer in both meters per second and mph.

---

[8] We'll assume the car has four wheels (and therefore four tires on the road) and that the mass of the car is evenly distributed among all four tires. Although you could draw four separate FBDs for the four tires, you should be able to convince yourself it is sufficient to consider the car and tires as one composite object. (Even though we didn't explicitly discuss it at the time, we did something similar every time we analyzed a cart moving on a track!)

**d.** Does the maximum speed change if you increase the mass of the car? Explain briefly.

You may have noticed that racetracks are often "banked." The next activity explores this situation in detail.

### 7.13.2. Activity: Don't Bank on it

Now suppose the racetrack in the previous problem is "banked" at a 25° angle so that the inner edge of the track is lower than the outer edge of the track (the road is tilted). You can assume everything else is the same as in the previous problem.

We'll let you tackle this problem following our usual problem-solving approach, starting with a FBD of the car as viewed from the rear. **Hint:** Even though the track is now tilted, it is far easier to keep your coordinate system oriented the same way as in the previous problem, with one axis aligned *toward the center of the circle* (you'll want to think carefully about the direction of the acceleration!).

**a.** Determine the *maximum* speed that the car can travel on this track without sliding off (assume it moves in a perfect circle at a constant speed). Give your answer in both meters per second and mph.

**b.** Compare your answer to the previous problem. Is this what you expected? Does this help explain why curves on a racetrack are banked?

**c.** Does the maximum speed change if you increase the mass of the car?

# APPENDIX A
# FURTHER DETAILS ON STATISTICS AND UNCERTAINTY

In Unit 2 we discussed concepts associated with measurement uncertainty, including inherent uncertainty, systematic errors, significant figures, standard deviation, histograms, standard deviation of the mean (SDM), and confidence intervals. Here we provide some additional details on related topics that were not discussed in Unit 2.

## HANDLING OF SIGNIFICANT FIGURES IN CALCULATIONS

When performing calculations the correct number of significant figures is obtained via propagation of uncertainty (see below for more details). However, such an analysis takes time and is frequently not practical during actual laboratory experimentation. In such cases, the following "rules of thumb" are often followed.

**Rule 1:** In multiplication or division, it is often acceptable to keep the same number of significant figures in the product or quotient as are in the *least* precise factor. In the examples that follow, the least precise factor has two significant figures, so the final answer is reported with two significant figures. Note that you should keep more than two significant figures during the calculation and only round at the end.

**Examples of Rule 1 (Multiplication and Division):**

$$3.6 \times 25.7 = 92.52 = 93$$

$$\frac{4.6}{757} = 0.006077 = 0.0061$$

**Rule 2:** In addition or subtraction, it is often acceptable to keep the same number of decimal places in the sum or difference as are in the number with the *least* number of decimal places. Again, you should keep additional significant figures and then round at the end.

**Examples of Rule 2 (Sum and Difference):**

| 37.6 | 6953 | 22.7 | 1.3378 |
|------|------|------|--------|
| − 2.45 | − 42.7 | 19.51 | 15.43 |
| 35.2 | 6910 | +.732 | +1.821 |
| | | 42.9 | 18.59 |

In general, one should retain enough significant figures that round-off error does not cause problems, but not so many as to constitute a burden. Here are a couple of examples where simply following the rules above leads to problems.

WRONG:   $0.77 \times 1.46 = 1.1$

In this case, following Rule 1 causes an issue. The numbers 0.77 and 1.46 are known to be precise to about 0.5%, whereas the incorrectly reported result of 1.1 is only precise to about 5%. In this extreme case, the precision of the result is reduced by almost a factor of ten due to round-off error. Thus, one should keep an additional significant figure in the answer, 1.12, but no more.

WRONG:   $0.77 \times 1.46 = 1.1242$

The extra digits, which are not actually significant, are just a burden. In addition, they carry the incorrect implication of a result with an extreme level of accuracy.

## PROPAGATION OF UNCERTAINTY

Suppose you measure a quantity and estimate its uncertainty. Now imagine you need to use the value you just measured to calculate a new quantity. Is there a way of determining the uncertainty of the new quantity based on the uncertainty in the original quantity? The answer is yes, and we do so by *propagating the uncertainty* (this is sometimes referred to as *error propagation*). Although we will not worry about propagating uncertainties in this course, it is important to know that there is a well-defined procedure through which one can calculate new uncertainties in terms of known uncertainties. We state here, without proof, a useful formula for propagating uncertainties in science and engineering for variables in which the errors are uncorrelated (which is nearly always the case in the physical sciences).

Assume we make a number of independent measurements $x_i$, where $i$ runs from 1 to $N$, and we want to calculate a new quantity that depends on these measurements in some way and can be described by a function $f(x_1, x_2, \ldots, x_N)$. For example, if you measure the radius $r$ of a sphere, the function $f$ might represent the formula for the sphere's volume: $f(r) = V(r) = \frac{4}{3}\pi r^3$. If the uncertainty of each measurement $x_i$ is given by $\sigma_i$, then the uncertainty in the calculated quantity $\sigma_f$ obeys the formula,[1,2]

$$\sigma_f^2 = \left(\frac{\partial f}{\partial x_1}\sigma_1\right)^2 + \left(\frac{\partial f}{\partial x_2}\sigma_2\right)^2 + \cdots + \left(\frac{\partial f}{\partial x_N}\sigma_N\right)^2 \tag{A.1}$$

It is sometimes said that the uncertainties are combined by *summation in quadrature*. Continuing with our example from above, if the uncertainty

in our radius measurement is given by $\sigma_r$, then the uncertainty in the volume, when calculated using this formula, is given by $\sigma_V = 4\pi r^2 \sigma_r$ (in this example there is only a single term on the right side of Eq. (A.1)). Notice that the uncertainty in the radius gets multiplied by the factor $4\pi r^2$ (the surface area of the sphere) to determine the uncertainty in the volume. Thus, we see that if the radius is large, a small uncertainty in its value can lead to a large uncertainty in the volume.

### Using Propagation of Uncertainty to Derive the SDM Formula

As we have discussed, the average, or mean, of a set of measurements is given by

$$\langle x \rangle \equiv \frac{x_1 + x_2 + \cdots + x_N}{N} = \frac{1}{N}\sum_{i=1}^{N} x_i$$

Thus, the mean is given by the function $\langle x \rangle = \frac{x_1}{N} + \frac{x_2}{N} + \cdots + \frac{x_N}{N}$. If we think of the mean as a function of a set of variables, $\langle x \rangle = f(x_1, x_2, \ldots, x_N)$, then we can use Eq. (A.1) to calculate the uncertainty in the average $\langle x \rangle$. To do so, we need the partial derivatives of the function $f$ with respect to each of the variables. Thankfully, the simple form of the function for the average makes this very easy, since each term only depends on one variable:

$$\frac{\partial f}{\partial x_1} = \frac{\partial}{\partial x_1}\left(\frac{x_1}{N} + \frac{x_2}{N} + \cdots + \frac{x_N}{N}\right) = \frac{1}{N}$$

Similarly, because each term has exactly the same form, every partial derivative will be exactly the same: $\frac{\partial f}{\partial x_2} = \frac{1}{N}$, $\frac{\partial f}{\partial x_3} = \frac{1}{N}$, etc. Now, because each of these measurements $x_i$ has the same uncertainty (which we take to be the standard deviation), we have $\sigma_1 = \sigma_2 = \cdots = \sigma_N \equiv \sigma_{sd}$. Thus, using Eq. (A.1) we find the uncertainty in the average to be

$$\sigma_{\langle x \rangle}^2 = \left(\frac{\partial f}{\partial x_1}\sigma_1\right)^2 + \left(\frac{\partial f}{\partial x_2}\sigma_2\right)^2 + \cdots + \left(\frac{\partial f}{\partial x_N}\sigma_N\right)^2$$

$$= \left(\frac{1}{N}\sigma_{sd}\right)^2 + \left(\frac{1}{N}\sigma_{sd}\right)^2 + \cdots + \left(\frac{1}{N}\sigma_{sd}\right)^2$$

$$= N\left(\frac{\sigma_{sd}^2}{N^2}\right) = \frac{\sigma_{sd}^2}{N}$$

[1] See, for example, Philip R. Bevington, *Data Reduction and Error Analysis for the Physical Sciences*, chapter 4 (McGraw-Hill, New York, 1969).
[2] The funny derivative symbols $\partial$ that appear in this formula represent *partial* derivatives. A partial derivative just means that all other variables are to be held constant when you differentiate with respect to the specified variable (you may have seen a similar formula in a multivariable calculus class). If this formula frightens you, rest assured that this appendix is the only time we will discuss partial derivatives in this class!

Finally, taking the square root of both sides leads to the expected result for the SDM:

$$\sigma_{\langle x \rangle} = \frac{\sigma_{sd}}{\sqrt{N}} \qquad (A.2)$$

## FITTING VIA THE METHOD OF LEAST SQUARES

In many of the activities in this Guide, you are asked to perform a "fit" to some data. The general idea is to find a mathematical function that describes the data in hopes of comparing the data to a model equation from some theory. If the relationship is simple enough, you could perform a fit "by eye." For example, if the data obviously lies along a straight line, you could make an estimate of the slope and $y$-intercept of a line you draw on a graph.

In general (and certainly for more complicated relationships), one uses a computer to perform a best fit to the data. The most common way of fitting a curve to the data is called the *method of least squares*, where the criterion for "best" fit is stated as follows: the sum of the squares of the vertical "distances" between each data point and the best-fit curve or line is a *minimum*. Most software packages for visualizing and analyzing data have built-in routines for performing a least-squares fit to a variety of different functions (linear, quadratic, exponential, sinusoidal, etc.). In this appendix we briefly outline the process of performing a linear least-squares fit to a set of data so that you have some understanding of how the fitting routine works.

### Least-Squares Fit for Linear Data

In the most elementary applications of the least-squares method, one of the coordinates of each point is assumed to be known exactly, so that it is entirely free from uncertainty (note that this can sometimes be a poor assumption!). The "uncertainty free" coordinate is typically chosen as the $x$-coordinate, and the experimental uncertainty in each data point $(x_i, y_i)$ resides only in $y_i$. Figure A.1 shows a set of experimental data points through which a least-squares straight line has been drawn.

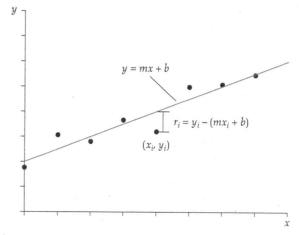

Fig. A.1. Graph showing a set of data points along with the best-fit line as determined by a linear least-squares fit.

The difference between a measured value $y_i$ and the height of the line $y = mx + b$ is defined to be the *residual* $r_i$. Therefore, each residual can be computed as

$$r_i = y_i - (mx_i + b) \qquad (A.3)$$

We want to find the slope $m$ and the $y$-intercept $b$ for which the sum of the squares of the residuals for all $n$ data points, $S$, is a minimum. That is, we select the values of $m$ and $b$ for which

$$S \equiv \sum_{i=1}^{n} r_i^2 = \text{a minimum} \qquad (A.4)$$

To find the best-fit values of $m$ and $b$, we begin by substituting Eq. (A.3) into Eq. (A.4) and expanding the square

$$S = \sum_{i=1}^{n} r_i^2 = \sum_{i=1}^{n} \left[ y_i - (mx_i + b) \right]^2$$

$$= \sum_{i=1}^{n} \left( y_i^2 + m^2 x_i^2 + b^2 + 2mbx_i - 2mx_i y_i - 2by_i \right)$$

Next, we want to find the values of $m$ and $b$ that *minimize* $S$, so we require $\partial S / \partial m = 0$ and $\partial S / \partial b = 0$. Recalling that the values of $x_i$ and $y_i$ represent specific values and are therefore constants for this problem, we find

$$\frac{\partial S}{\partial m} = 2m \sum_{i=1}^{n} x_i^2 + 2b \sum_{i=1}^{n} x_i - 2 \sum_{i=1}^{n} x_i y_i = 0$$

and

$$\frac{\partial S}{\partial b} = 2nb + 2m\sum_{i=1}^{n} x_i - 2\sum_{i=1}^{n} y_i = 0$$

These last two equations can then be solved for $m$ and $b$, yielding

$$m = \frac{n\sum x_i y_i - \sum x_i \sum y_i}{n\sum x_i^2 - \left(\sum x_i\right)^2} \qquad (A.5)$$

$$b = \frac{\sum x_i^2 \sum y_i^2 - \sum x_i \sum x_i y_i}{n\sum x_i^2 - \left(\sum x_i\right)^2} \qquad (A.6)$$

where all sums run from $i = 1$ to $i = N$.

### Interpreting the Fit

Assuming a sufficiently large data set and normally distributed experimental uncertainty, the standard deviation represents approximately a 68% confidence interval. This indicates that the true value of the slope (and intercept) of the process from which the data set was taken has about a 68% chance of lying within one standard deviation of the slope (and intercept) found by the least-squares analysis.

The resulting fit will typically report the value of the square of the *correlation coefficient*, which is an important indicator of the validity of your fit equation. The correlation coefficient reflects the proportion of the variations in the measured data values about their means that are attributable to the experimental uncertainty, which is assumed to be normally distributed. The remaining portion of the variation in the measured data is due to the inability of the fit equation to accurately describe the data. Therefore, as the value of the correlation coefficient approaches one, all of the variation in the data can be attributed to experimental uncertainty, and the fit equation is considered to be an accurate model of the data. However, as the correlation coefficient decreases below one, more of the variations are due to the fact that the data are not accurately described by the fit equation.

### FURTHER READING

Two of the more informative books on measurement uncertainties are:

Baird, D.C., *Experimentation: An Introduction to Measurement Theory and Experiment Design* (Prentice-Hall, Englewood Cliffs, NJ, 1988).

Taylor, J.R., *An Introduction to Error Analysis: The Study of Uncertainties in Physical Measurements* (University Science Books, Mill Valley, CA, 1982).

# APPENDIX B
# NOTATION AND REFERENCE TABLES

Poorly chosen mathematical notation can be a source of considerable confusion for those trying to learn and master physics. For example, ambiguity in the meaning of a mathematical symbol can prevent a reader from understanding the meaning of a crucial relationship. It is also difficult to solve problems when the symbols used to represent different quantities are not distinctive. In this text, we have done our best to use mathematical notation in ways that allow important distinctions to be easily visible both on the printed page and in handwritten work. We include below a number of tables that provide important notes on notation, physical constants, unit conversions, etc. Some of these tables are reproduced in the front and back covers for easy reference while working through the activities.

## PHYSICAL PROPERTIES

Air (at room temperature and sea level atmospheric pressure)

| | |
|---|---|
| Density | $1.20 \, \text{kg/m}^3$ |
| Specific heat at constant pressure ($C_p$) | $1.00 \times 10^3 \, \text{J kg}^{-1} \, \text{K}^{-1}$ |
| Speed of sound | $343 \, \text{m/s}$ |

Water (at room temperature and sea level atmospheric pressure)

| | |
|---|---|
| Density | $1.00 \times 10^3 \, \text{kg/m}^3$ |
| Specific heat | $4.18 \times 10^3 \text{J kg}^{-1} \, \text{K}^{-1}$ |
| Speed of sound | $1.26 \times 10^3 \, \text{m/s}$ |

Earth

| | |
|---|---|
| Density (mean) | $5.49 \times 10^3 \, \text{kg/m}^3$ |
| Radius (mean) | $6.37 \times 10^6 \, \text{m}$ |
| Mass | $5.97 \times 10^{24} \, \text{kg}$ |
| Atmospheric pressure (average sea level) | $1.01 \times 10^5 \, \text{Pa}$ |
| Mean Earth-moon distance | $3.84 \times 10^8 \, \text{m}$ |

## SOLAR SYSTEM

| Body | Mean radius of orbit (m) | Mean radius of body (m) | Mass (kg) |
|---|---|---|---|
| Sun | | $6.96 \times 10^8$ | $1.99 \times 10^{30}$ |
| Mercury | $5.79 \times 10^{10}$ | $2.42 \times 10^6$ | $3.35 \times 10^{23}$ |
| Venus | $1.08 \times 10^{11}$ | $6.10 \times 10^6$ | $4.89 \times 10^{24}$ |
| Earth | $1.50 \times 10^{11}$ | $6.37 \times 10^6$ | $5.97 \times 10^{24}$ |
| Mars | $2.28 \times 10^{11}$ | $3.38 \times 10^6$ | $6.46 \times 10^{23}$ |
| Jupiter | $7.78 \times 10^{11}$ | $7.13 \times 10^7$ | $1.90 \times 10^{27}$ |
| Saturn | $1.43 \times 10^{12}$ | $6.04 \times 10^7$ | $5.69 \times 10^{26}$ |
| Moon | $3.84 \times 10^8$ | $1.74 \times 10^6$ | $7.35 \times 10^{22}$ |

## SI MULTIPLIERS

| Abbreviation | Name | Value |
|---|---|---|
| Y | yotta | $10^{24}$ |
| Z | zetta | $10^{21}$ |
| E | exa | $10^{18}$ |
| P | peta | $10^{15}$ |
| T | tera | $10^{12}$ |
| G | giga | $10^9$ |
| M | mega | $10^6$ |
| k | kilo | $10^3$ |
| c | centi | $10^{-2}$ |
| m | milli | $10^{-3}$ |
| μ | micro | $10^{-6}$ |
| n | nano | $10^{-9}$ |
| p | pico | $10^{-12}$ |
| f | femto | $10^{-15}$ |
| a | atto | $10^{-18}$ |
| z | zepto | $10^{-21}$ |
| y | yocto | $10^{-24}$ |

## PHYSICAL CONSTANTS

| | | |
|---|---|---|
| Local gravitational field strength (near Earth) | $g$ | 9.81 N/kg |
| Gravitational constant | $G$ | $6.67 \times 10^{-11}$ N m$^2$/kg$^2$ |
| Electron mass | $m_e$ | $9.11 \times 10^{-31}$ kg |
| Proton mass | $m_p$ | $1.673 \times 10^{-27}$ kg |
| Neutron mass | $m_n$ | $1.675 \times 10^{-27}$ kg |
| Speed of light | $c$ | $3.00 \times 10^8$ m/s |
| Universal gas constant | $R$ | 8.31 J mol$^{-1}$ K$^{-1}$ |
| Boltzmann's constant | $k_B$ | $1.38 \times 10^{-23}$ J/K |
| Avogadro's number | $N_A$ | $6.02 \times 10^{23}$/mol |
| Electric constant (permittivity) | $\varepsilon_0$ | $8.85 \times 10^{-12}$ F/m |
| Coulomb constant | $k = \dfrac{1}{4\pi\varepsilon_0}$ | $8.99 \times 10^9$ N m$^2$/C$^2$ |
| Elementary charge | $e$ | $1.60 \times 10^{-19}$ C |
| Magnetic constant (permeability) | $\mu_0$ | $4\pi \times 10^{-7}$ T m/A |
| Unified atomic mass unit | u | $1.66 \times 10^{-27}$ kg |

## CONVERSIONS

Length

- 1 in = 2.54 cm
- 1 ft = 12 in = 0.3048 m
- 1 m = 39.37 in = 3.281 ft
- 1 yd = 3 ft = 0.9144 m
- 1 km = 0.621 mi
- 1 mi = 1.609 km = 5280 ft
- 1 lightyear = $9.461 \times 10^{15}$ m
- 1 Å = $10^{-10}$ m

Area

- 1 m$^2$ = $10^4$ cm$^2$ = 10.76 ft$^2$
- 1 ft$^2$ = 0.0929 m$^2$ = 144 in$^2$
- 1 in$^2$ = 6.452 cm$^2$

Volume

- 1 m$^3$ = $10^6$ cm$^3$ = 35.3 ft$^3$
- 1 ft$^3$ = $2.83 \times 10^{-2}$ m$^3$ = 1728 in$^3$
- 1 liter = 1000 cm$^3$ = 1.0576 qt = 0.0353 ft$^3$
- 1 ft$^3$ = 7.481 gal = 28.32 liters
- 1 gal = 3.786 liters = 231 in$^3$

Pressure

- 1 Pa = 1 N/m$^2$ = $1.45 \times 10^{-4}$ lb/in$^2$
- 1 atm = $1.013 \times 10^5$ Pa = 14.7 lb/in$^2$ = 760 mm Hg
- 1 bar = $10^5$ Pa = 14.50 lb/in$^2$

Mass

- 1000 kg = 1 t (metric ton)
- 1 slug = 14.59 kg
- 1 u = $1.66 \times 10^{-27}$ kg = 931.5 MeV/c$^2$

Force

- 1 N = 0.2248 lb
- 1 lb = 4.448 N

Velocity

- 1 m/s = 3.28 ft/s = 2.24 mi/hr
- 1 mi/hr = 1.61 km/hr = 0.447 m/s = 1.47 ft/s
- 1 mi/min = 60 mi/hr = 88 ft/s

Acceleration

- 1 m/s$^2$ = 3.28 ft/s$^2$
- 1 ft/s$^2$ = 0.3048 m/s$^2$ = 30.48 cm/s$^2$

Time

- 1 day = 24 hr = $1.44 \times 10^3$ min = $8.64 \times 10^4$ s
- 1 year = 365 days = $3.16 \times 10^7$ s

Energy

- 1 J = 1 N m = 0.738 ft lb = $10^7$ ergs
- 1 cal = 4.186 J
- 1 Btu = 252 cal = $1.054 \times 10^3$ J
- 1 eV = $1.6 \times 10^{-19}$ J
- 1 kWh = $3.60 \times 10^6$ J

Power

- 1 hp = 0.746 kW = 550 ft·lb/s
- 1 W = 1 J/s = 0.738 ft·lb/s
- 1 Btu/hr = 0.293 W

## UNITS

| Quantity | Unit | In terms of base units | In other common units |
|---|---|---|---|
| Capacitance ($C$) | farad (F) | $kg^{-1} \cdot m^{-2} \cdot s^4 \cdot A^2$ | C/V |
| Electric charge ($q$) | coulomb (C) | $s \cdot A$ | |
| Electric field ($E$) | | $kg \cdot m \cdot s^{-3} \cdot A^{-1}$ | N/C or V/m |
| Electric potential ($V$) (also emf [$\varepsilon$]) | volt (V) | $kg \cdot m^2 \cdot s^{-3} \cdot A^{-1}$ | J/C or W/A |
| Electric resistance ($R$) | ohm ($\Omega$) | $kg \cdot m^2 \cdot s^{-3} \cdot A^{-2}$ | V/A |
| Energy ($E$) | joule (J) | $kg \cdot m^2/s^2$ | N m |
| Force ($F$) | newton (N) | $kg \cdot m/s^2$ | |
| Frequency ($f$) | hertz (Hz) | $s^{-1}$ | |
| Magnetic field ($B$) | tesla (T) | $kg \cdot s^{-2} \cdot A^{-1}$ | Wb/m$^2$ |
| Magnetic flux ($\Phi^{mag}$) | weber (Wb) | $kg \cdot m^2 \cdot s^{-2} \cdot A^{-1}$ | V s |
| Power ($P$) | watt (W) | $kg \cdot m^2/s^3$ | J/s |
| Pressure ($P$) | pascal (Pa) | $kg \cdot m^{-1} \cdot s^{-2}$ | N/m$^2$ or J/m$^3$ |

## SYMBOLS USED IN THIS ACTIVITY GUIDE

| | |
|---|---|
| $\vec{a}$ | acceleration |
| $A$ | area |
| $\vec{B}$ | magnetic field |
| $c$ | specific heat (per unit mass); speed of light |
| $C_p$ | molar specific heat at constant pressure |
| $C_V$ | molar specific heat at constant volume |
| $C$ | capacitance |
| $e$ | magnitude of charge of electron; base of natural logs, 2.71828 … |
| $E$ | energy |
| $\vec{E}$ | electric field |
| $\varepsilon$ | emf (electromotive force) |
| $\varepsilon_0$ | permittivity of free space |
| $f$ | frequency; number of degrees of freedom |
| $\vec{f}$ | frictional force (static or kinetic) |
| $\vec{F}$ | force |
| $g$ | local gravitational field strength |
| $G$ | gravitational constant |
| $i$ | current (assumed to be direction of positive charge carriers) |
| $I$ | rotational inertia |
| $\vec{J}$ | impulse |
| $k_B$ | Boltzmann's constant |
| $k$ | Coulomb's constant; spring constant |
| $K$ | kinetic energy |
| $L$ | latent heat |
| $\vec{\ell}, \vec{L}$ | rotational momentum |
| $m, M$ | mass |
| $n$ | number of moles; number of density of charge carriers; number of turns per unit length |
| $N$ | neutron number of nucleus; number of turns in a coil; total number of particles |
| $\vec{p}, \vec{P}$ | momentum |

| | |
|---|---|
| $P$ | power; pressure |
| $q$ | charge of particle |
| $Q$ | heat (thermal energy transfer); charge of system |
| $\vec{r}$ | position vector |
| $R$ | universal gas constant; resistance |
| $S$ | entropy |
| $t$ | time |
| $T$ | temperature; period |
| $U$ | potential energy |
| $v$ | speed |
| $\vec{v}$ | velocity |
| $\Delta V$ | electric potential difference |
| $W$ | work |
| $\hat{x}, \hat{y}, \hat{z}$ | unit vectors in the $x, y, z$ directions |
| $Z$ | atomic number of nucleus |
| $\alpha$ | angular acceleration (magnitude); alpha particle (He nucleus) |
| $\beta$ | beta particle (electron) |
| $\gamma$ | ratio of specific heats, $C_p/C_V$ |
| $\theta$ | angle; rotational position |
| $\kappa$ | dielectric constant |
| $\lambda$ | wavelength; decay constant; charge per unit length |
| $\mu$ | coefficient of friction (static or kinetic); mass per unit length; permeability |
| $\rho$ | mass (or charge) per unit volume; resistivity |
| $\sigma$ | charge per unit area; conductivity |
| $\tau$ | torque (magnitude); time constant; mean time between collisions |
| $\vec{\tau}$ | torque |
| $\Phi$ | flux of a vector field |
| $\phi$ | phase constant |
| $\omega$ | rotational speed; rotational frequency |
| $\vec{\omega}$ | rotational velocity |

## VECTOR PRODUCTS

$$\vec{A} = A_x \hat{x} + A_y \hat{y} + A_z \hat{z}$$

$$\vec{B} = B_x \hat{x} + B_y \hat{y} + B_z \hat{z}$$

$$\vec{A} \cdot \vec{B} = A_x B_x + A_y B_y + A_z B_z$$

$$\vec{A} \times \vec{B} = \begin{vmatrix} \hat{x} & \hat{y} & \hat{z} \\ A_x & A_y & A_z \\ B_x & B_y & B_z \end{vmatrix} = \begin{matrix} (A_y B_z - A_z B_y)\hat{x} \\ -(A_x B_z - A_z B_x)\hat{y} \\ +(A_x B_y - A_y B_x)\hat{z} \end{matrix}$$

## THE GREEK ALPHABET

| Alpha | A α | Eta | H η | Nu | N ν | Tau | T τ |
| Beta | B β | Theta | Θ θ | Xi | Ξ ξ | Upsilon | Υ υ |
| Gamma | Γ γ | Iota | I ι | Omicron | O o | Phi | Φ φ |
| Delta | Δ δ | Kappa | K κ | Pi | Π π | Chi | X χ |
| Epsilon | E ε | Lambda | Λ λ | Rho | P ρ | Psi | Ψ ψ |
| Zeta | Z ζ | Mu | M μ | Sigma | Σ σ | Omega | Ω ω |

## USEFUL NUMBERS + TRIGONOMETRY

$\pi = 3.14159$
$e = 2.71828$

$1 \text{ rad} = 57.2958°$

$\ln 2 = 0.693147$
$\ln 10 = 2.30259$

$\sin 0 = 0$
$\cos 0 = 1$
$\tan 0 = 0$

$\sin 30° = \dfrac{1}{2}$

$\cos 30° = \dfrac{\sqrt{3}}{2}$

$\tan 30° = \dfrac{1}{\sqrt{3}}$

$\left(30° = \dfrac{\pi}{6}\text{rad}\right)$

$\sin 45° = \dfrac{\sqrt{2}}{2}$

$\cos 45° = \dfrac{\sqrt{2}}{2}$

$\tan 45° = 1$

$\left(45° = \dfrac{\pi}{4}\text{rad}\right)$

$\sin 60° = \dfrac{\sqrt{3}}{2}$

$\cos 60° = \dfrac{1}{2}$

$\tan 60° = \sqrt{3}$

$\left(60° = \dfrac{\pi}{3}\text{rad}\right)$

$\sin 90° = 1$

$\cos 90° = 0$

$\tan 90° = \infty$

$\left(90° = \dfrac{\pi}{2}\text{rad}\right)$

## TRIGONOMETRIC FUNCTIONS AND THE UNIT CIRCLE

$$\sin \theta = \frac{y}{r}$$

$$\cos \theta = \frac{x}{r}$$

$$\tan \theta = \frac{\sin \theta}{\cos \theta} = \frac{y}{x}$$

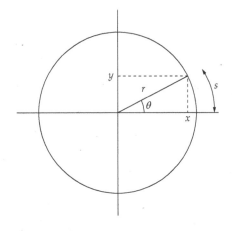

Definition of angle (in radians): $\theta = \dfrac{s}{r}$

$2\pi$ radians in complete circle
1 radian = 57.3°

# INDEX

## SYMBOLS USED IN THIS ACTIVITY GUIDE

| | |
|---|---|
| $\vec{a}$ | acceleration |
| $A$ | area |
| $\vec{B}$ | magnetic field |
| $c$ | specific heat (per unit mass); speed of light |
| $C_p$ | molar specific heat at constant pressure |
| $C_V$ | molar specific heat at constant volume |
| $C$ | capacitance |
| $e$ | magnitude of charge of electron; base of natural logs, 2.71828… |
| $E$ | energy |
| $\vec{E}$ | electric field |
| $\varepsilon$ | emf (electromotive force) |
| $\varepsilon_0$ | permittivity of free space |
| $f$ | frequency; number of degrees of freedom |
| $\vec{f}$ | frictional force (static or kinetic) |
| $\vec{F}$ | force |
| $g$ | local gravitational field strength |
| $G$ | gravitational constant |
| $i$ | current (assumed to be direction of positive charge carriers) |
| $I$ | rotational inertia |
| $\vec{J}$ | impulse |
| $k_B$ | Boltzmann's constant |
| $k$ | Coulomb's constant; spring constant |
| $K$ | kinetic energy |
| $L$ | latent heat |
| $\vec{\ell}, \vec{L}$ | rotational momentum |
| $m, M$ | mass |
| $n$ | number of moles; number of density of charge carriers; number of turns per unit length |
| $N$ | neutron number of nucleus; number of turns in a coil; total number of particles |
| $\vec{p}, \vec{P}$ | momentum |

| | |
|---|---|
| $P$ | power; pressure |
| $q$ | charge of particle |
| $Q$ | heat (thermal energy transfer); charge of system |
| $\vec{r}$ | position vector |
| $R$ | universal gas constant; resistance |
| $S$ | entropy |
| $t$ | time |
| $T$ | temperature; period |
| $U$ | potential energy |
| $v$ | speed |
| $\vec{v}$ | velocity |
| $\Delta V$ | electric potential difference |
| $W$ | work |
| $\hat{x}, \hat{y}, \hat{z}$ | unit vectors in the $x, y, z$ directions |
| $Z$ | atomic number of nucleus |
| $\alpha$ | angular acceleration (magnitude); alpha particle (He nucleus) |
| $\beta$ | beta particle (electron) |
| $\gamma$ | ratio of specific heats, $C_p/C_V$ |
| $\theta$ | angle; rotational position |
| $\kappa$ | dielectric constant |
| $\lambda$ | wavelength; decay constant; charge per unit length |
| $\mu$ | coefficient of friction (static or kinetic); mass per unit length; permeability |
| $\rho$ | mass (or charge) per unit volume; resistivity |
| $\sigma$ | charge per unit area; conductivity |
| $\tau$ | torque (magnitude); time constant; mean time between collisions |
| $\vec{\tau}$ | torque |
| $\Phi$ | flux of a vector field |
| $\phi$ | phase constant |
| $\omega$ | rotational speed; rotational frequency |
| $\vec{\omega}$ | rotational velocity |

## PHYSICAL CONSTANTS

| | | |
|---|---|---|
| Local gravitational field strength (near Earth) | $g$ | 9.81 N/kg |
| Gravitational constant | $G$ | $6.67 \times 10^{-11}$ N m$^2$/kg$^2$ |
| Electron mass | $m_e$ | $9.11 \times 10^{-31}$ kg |
| Proton mass | $m_p$ | $1.673 \times 10^{-27}$ kg |
| Neutron mass | $m_n$ | $1.675 \times 10^{-27}$ kg |
| Speed of light | $c$ | $3.00 \times 10^8$ m/s |
| Universal gas constant | $R$ | 8.31 J mol$^{-1}$ K$^{-1}$ |
| Boltzmann's constant | $k_B$ | $1.38 \times 10^{-23}$ J/K |
| Avogadro's number | $N_A$ | $6.02 \times 10^{23}$/mol |
| Electric constant (permittivity) | $\varepsilon_0$ | $8.85 \times 10^{-12}$ F/m |
| Coulomb constant | $k = \dfrac{1}{4\pi\varepsilon_0}$ | $8.99 \times 10^9$ N m$^2$/C$^2$ |
| Elementary charge | $e$ | $1.60 \times 10^{-19}$ C |
| Magnetic constant (permeability) | $\mu_0$ | $4\pi \times 10^{-7}$ T m/A |
| Unified atomic mass unit | u | $1.66 \times 10^{-27}$ kg |

## PHYSICAL PROPERTIES

Air (at room temperature and sea level atmospheric pressure)

| | |
|---|---|
| Density | 1.20 kg/m$^3$ |
| Specific heat at constant pressure ($C_p$) | $1.00 \times 10^3$ J kg$^{-1}$ K$^{-1}$ |
| Speed of sound | 343 m/s |

Water (at room temperature and sea level atmospheric pressure)

| | |
|---|---|
| Density | $1.00 \times 10^3$ kg/m$^3$ |
| Specific heat | $4.18 \times 10^3$ J kg$^{-1}$ K$^{-1}$ |
| Speed of sound | $1.26 \times 10^3$ m/s |

Earth

| | |
|---|---|
| Density (mean) | $5.49 \times 10^3$ kg/m$^3$ |
| Radius (mean) | $6.37 \times 10^6$ m |
| Mass | $5.97 \times 10^{24}$ kg |
| Atmospheric pressure (average sea level) | $1.01 \times 10^5$ Pa |
| Mean Earth-moon distance | $3.84 \times 10^8$ m |

## CONVERSIONS

### Length

1 in = 2.54 cm

1 ft = 12 in = 0.3048 m

1 m = 39.37 in = 3.281 ft

1 yd = 3 ft = 0.9144 m

1 km = 0.621 mi

1 mi = 1.609 km = 5280 ft

1 lightyear = $9.461 \times 10^{15}$ m

1 Å = $10^{-10}$ m

### Area

1 m$^2$ = $10^4$ cm$^2$ = 10.76 ft$^2$

1 ft$^2$ = 0.0929 m$^2$ = 144 in$^2$

1 in$^2$ = 6.452 cm$^2$

### Volume

1 m$^3$ = $10^6$ cm$^3$ = 35.3 ft$^3$

1 ft$^3$ = $2.83 \times 10^{-2}$ m$^3$ = 1728 in$^3$

1 liter = 1000 cm$^3$ = 1.0576 qt = 0.0353 ft$^3$

1 ft$^3$ = 7.481 gal = 28.32 liters

1 gal = 3.786 liters = 231 in$^3$

### Pressure

1 Pa = 1 N/m$^2$ = $1.45 \times 10^{-4}$ lb/in$^2$

1 atm = $1.013 \times 10^5$ Pa = 14.7 lb/in$^2$ = 760 mm Hg

1 bar = $10^5$ Pa = 14.50 lb/in$^2$

### Mass

1000 kg = 1 t (metric ton)

1 slug = 14.59 kg

1 u = $1.66 \times 10^{-27}$ kg = 931.5 MeV/c$^2$

### Force

1 N = 0.2248 lb

1 lb = 4.448 N

### Velocity

1 m/s = 3.28 ft/s = 2.24 mi/hr

1 mi/hr = 1.61 km/hr = 0.447 m/s = 1.47 ft/s

1 mi/min = 60 mi/hr = 88 ft/s

### Acceleration

1 m/s$^2$ = 3.28 ft/s$^2$

1 ft/s$^2$ = 0.3048 m/s$^2$ = 30.48 cm/s$^2$

### Time

1 day = 24 hr = $1.44 \times 10^3$ min = $8.64 \times 10^4$ s

1 year = 365 days = $3.16 \times 10^7$ s

### Energy

1 J = 1 N m = 0.738 ft lb = $10^7$ ergs

1 cal = 4.186 J

1 Btu = 252 cal = $1.054 \times 10^3$ J

1 eV = $1.6 \times 10^{-19}$ J

1 kWh = $3.60 \times 10^6$ J

### Power

1 hp = 0.746 kW = 550 ft·lb/s

1 W = 1 J/s = 0.738 ft·lb/s

1 Btu/hr = 0.293 W

## SOLAR SYSTEM

| Body | Mean radius of orbit (m) | Mean radius of body (m) | Mass (kg) |
|------|--------------------------|--------------------------|-----------|
| Sun | | $6.96 \times 10^8$ | $1.99 \times 10^{30}$ |
| Mercury | $5.79 \times 10^{10}$ | $2.42 \times 10^6$ | $3.35 \times 10^{23}$ |
| Venus | $1.08 \times 10^{11}$ | $6.10 \times 10^6$ | $4.89 \times 10^{24}$ |
| Earth | $1.50 \times 10^{11}$ | $6.37 \times 10^6$ | $5.97 \times 10^{24}$ |
| Mars | $2.28 \times 10^{11}$ | $3.38 \times 10^6$ | $6.46 \times 10^{23}$ |
| Jupiter | $7.78 \times 10^{11}$ | $7.13 \times 10^7$ | $1.90 \times 10^{27}$ |
| Saturn | $1.43 \times 10^{12}$ | $6.04 \times 10^7$ | $5.69 \times 10^{26}$ |
| Moon | $3.84 \times 10^8$ | $1.74 \times 10^6$ | $7.35 \times 10^{22}$ |

## VECTOR PRODUCTS

$$\vec{A} = A_x\hat{x} + A_y\hat{y} + A_z\hat{z}$$

$$\vec{B} = B_x\hat{x} + B_y\hat{y} + B_z\hat{z}$$

$$\vec{A} \cdot \vec{B} = A_xB_x + A_yB_y + A_zB_z$$

$$\vec{A} \times \vec{B} = \begin{vmatrix} \hat{x} & \hat{y} & \hat{z} \\ A_x & A_y & A_z \\ B_x & B_y & B_z \end{vmatrix} = \begin{matrix} (A_yB_z - A_zB_y)\hat{x} \\ -(A_xB_z - A_zB_x)\hat{y} \\ +(A_xB_y - A_yB_x)\hat{z} \end{matrix}$$